THE ECOLOGICAL BASIS
OF PLANNING

by

ARTUR GLIKSON

edited by

LEWIS MUMFORD

MARTINUS NIJHOFF / THE HAGUE / 1971

ISBN 90 247 1193 2

L C

PRINTED IN THE NETHERLANDS

THE ECOLOGICAL BASIS OF PLANNING

CONTENTS

INTRODUCTION, by Lewis Mumford   vii

    I. MAN'S RELATIONSHIP TO HIS ENVIRONMENT   1

    II. RECREATIONAL LAND USE   17

    III. PLANNING WITH THE LAND   36

    IV. THE PLANNER GEDDES   45

    V. SOCIAL TRENDS IN REGIONAL PLANNING   52

    VI. THE RELATIONSHIP BETWEEN LANDSCAPE PLANNING AND RE-
GIONAL PLANNING   67

    VII. TOWARDS REGIONAL LANDSCAPE DESIGN   79

    VIII. NOTES ON REGIONAL PLANNING AND TECHNOLOGICAL PRO-
GRESS   87

    IX. CREATING NEW LAND: DESIGNING ON NEW LAND   99

    X. THE INTEGRAL HABITATIONAL UNIT   110

INTRODUCTION

When Artur Glikson died in July 1966 he was still comparatively unknown; yet paradoxically he had an international reputation that went beyond town planning and architectural circles. As far back as 1955, when he was forty-four years old, he was an active participant in the notable Wenner-Gren Conference on "Man's Role in Changing the Face of the Earth," where he presented the first paper in the present book. Seven years later he was the only nonscientist represented in the even more selective Ciba Foundation conference on Man and his Future. Though Glikson attended many other important international conferences, notably the International Seminar on Regional Planning in The Hague in 1957, and the International conference of Landscape Architects in Amsterdam in 1960, he has yet to leave his mark on the thought and practice of architects and planners, his own professional group.

The fact that Artur Glikson's activities as a pioneer in sociological planning are still relatively unknown, might seem a handicap from the point of this book's getting the public or professional attention that it deserves. But this is perhaps the best reason for bringing out the assembled papers and giving a picture of their background in his personal experience. What makes Glikson's work important is that it represents in a series of concrete experiments and theoretic explorations a profound re-orientation that is taking place in many fields: a departure from the single-factor analysis, the mechanistic oversimplifications, and the unbalanced activities that are now undermining the very basis of organic existence, to say nothing of human culture. Glikson was concerned not only to overcome the ecological devastation produced by the dominant power system of the last four centuries: he sought to open up a far richer and more varied life than that promised by either sterile space adventures to distant planets unfit for human habitation, or by reducing human functions and needs to those that would ensure a precarious survival in a space capsule.

There is a reason for Glikson's still limited following. He lacked the qualities of an actor and a showman which, as in the case of Le Corbusier and Frank Lloyd Wright, give a man an early reputation .Glikson was too modest, too humorously self-critical, too sensitive, too respectful of life's complexities and difficulties to be tempted to cut such a figure. Above all he had a respect for the integrity of those whose work he valued: he obeyed Pascal's adminition not to speak of "my" ideas, as if ideas were personal property, but of *our* ideas. Men of this character often remain obscure in their own lifetime. In one of his last essays, Thomas Mann makes this observation about Chekhov, whom he confessed he himself had long undervalued, because the immediate impression he made lacked the dramatic flair of a Goethe or a Tolstoi. The modesty of such men is often treated as a sign of inferior talents: the world takes them at their own unassuming face value, not understanding the deeper pride and self-confidence that underlies such reticence. Such leaders must often wait quietly, for a generation or two if need be, before their quality is fully recognized.

Whatever Artur Glikson's misgivings from time to time about his own achievements as an architect, an urban planner, or a teacher, he had no such doubts about the body of ideas that he, perhaps more fully than any other members of his generation, represented and carried further. A few years before his death he went back to an idea that had tempted him even earlier: that of publishing possibly ten of his papers and articles in book form; to which he added hopefully a volume of photographs on men and cities and landscapes, which would put in graphic form relationships between man and his habitat that cannot be reduced to words. Some of these essays were gathered together, at least in part, in two memorial brochures, one in Hebrew, one in English, put out by the Israel Housing Ministry in 1967. So it is both out of friendship and out of our common concern for the ideas that Glikson expressed so cogently that I have felt it a duty to edit this volume, making only such minimal changes in his otherwise remarkably adequate and accurate English as were necessary to clarify his intention or to remove an occasional awkwardness.

Before attempting to present the intellectual background of Glikson's outlook and his specific planning ideas, let me first give a brief sketch of his life; for as I read the history of our times Glikson's life, while grievously handicapped and circumscribed by the worldwide cultural and moral disintegration of the last half-century, nevertheless was able to overcome in his own person the dehumanizing processes of our compulsive power-centered civilization, and to open up new prospects for both the community and the personality. Glikson himself was a living example of an emerging type of personality, one deeply rooted in particular landscapes and cultures, appreciating their unique potentialities, but open to contacts, cooperations, and trans-national conversations on a worldwide scale. In short, he was both a regionalist and a universalist: a fact really symbolized by this very book's being brought out by a Dutch publisher, with the backing of Israeli, German, French, English, Italian, Greek and Cretan colleagues, and an American editor. Thus he embodied, in his own person, the new conception of "One World Man." (See The Transformations of Man. New York: 1956.)

Artur Glikson was born at Koenigsberg, Germany, in 1911 and finished his secondary education in 1929. Afted receiving his diploma in architecture from the Technische Hochschule at Berlin-Charlottenburg, he emigrated in 1935 to Palestine. His independent career as architect and planner began in 1938 in Petah-Tiqva, a site not far from Tel-Aviv, where he prepared both the master plan and detailed schemes for squares, streets, and public buildings, such as schools and clinics. Since housing from 1948 on was the main problem for the newly founded state of Israel, open by necessity to tens of thousands of desperate refugees, this work was at first central to all his other activities. At the same time as a department head in the Planning Division, Glikson participated in the formulation of the National Plan and several regional plans. Such varied duties and constructive opportunities come all too rarely in a planner's life; and they gave Glikson's thought a concrete, down-to-earth quality that kept him from the technological fantasies of a Buckminster Fuller or the sterile esthetic audacities of a Le Corbusier or a Kenzo Tange. "Down to earth" in Glikson's thought always meant "close to man."

The severe constrictions of time and money, to say nothing of the pressure of immediate needs in planning new settlements and towns in Israel made any "ideal" solution impossible: such immediate planning at best, was preparatory and provisional. Yet this was an admirable post-graduate school for any planner, despite its handicaps; for it put every urban problem in its regional setting, as conditioned by climate, geological formation, soil, water, agricultural productivity, and industrial potentiality: so that the planner could never entertain the common metropolitan illusion that the city was in any sense a self-contained entity, nor could it continue to grow indefinitely, as metropolises so often do, without paying the faintest attention, until disaster threatens, to the effect of overgrowth on water pollution, sewage disposal, waste removal, and recreational opportunity, as well as upon the immediate quality of life in the congested inner core or over-dispersed suburban wasteland. When Glikson's planning unit was "re-organized" – read disbanded! – in 1953, it was because the new minister, the former Mayor of

Tel-Aviv, felt that National Planning was only a hindrance to cities like Tel-Aviv, which would limit its economic dynamism and prevent it from being a metropolis of a million inhabitants. This old-fashioned concept of the "great city," common to the one-time Mayor of Tel-Aviv, to Le Corbusier, and to Jane Jacobs, is precisely what Glikson's ecological concepts challenged. Fortunately, Glikson's earlier intellectual preparation had given his work a firm foundation.

While acquiring this planning experience, Glikson was compelled to live with his family, or sometimes alone, under minimal conditions of housing. These were cramped, embarrassing conditions, particularly for a mind as reflective as his, needing opportunity for withdrawal and privacy. But the very severity of such overcrowding brought its own kind of reward. Glikson was living in quarters whose domestic limitations were those under which the greater part of mankind still live, even in supposedly prosperous countries. So, while he had a keen appreciation of such esthetic creativity as Le Corbusier, for example, showed in his Marseille Unity House, he could never countenance Le Corbusier's highhanded way of overlooking exorbitant building costs, diverting funds from more essential human needs. Glikson understood the point of John Ruskin's remarks to the industrialists of Bradford, when Ruskin told them to forget about fine architecture until they had succeeded in purifying their foul air from smoke and their water from contamination and pollution. But along with this, Glikson appreciated the positive values of humble life.

For a professional whose life was immersed from the beginning in the practical details of planning, Glikson maintained, better than most active professionals, an intellectual life that covered a wide range of interests, from archaeology and history and religion on to biology and sociology. His earliest paper, to which a date can be affixed, is that on "Stabkirche und Wikingerschiff," in 1934; and there is another paper, perhaps earlier, on the relation of Marxian teaching to Man. Despite his activity as architect, he published "thoughts on Religion," in German, in 1941. In the range of these papers one has some insight into those human qualities that made Glikson always something more than a "specialist" or an "expert"; for he was above all, a thinking and feeling human being, at home not only with the objective methods of science and technics, but equally open to the subjective worlds of poetry and music.

Among Glikson's surviving papers, dating from the early nineteen-forties, there is not only a remarkably self-revealing diary, but a number of sensitive verses and prose poems, written in German. There are gnomic aphorisms, too, which I am tempted to quote. Let me give a few examples.

Es rieselt in Gärten das Wasser
Es rieselt um einen die Zeit,
Und niemand fragt dich:
"Bereit?"
und niemand sagt:
"Es ist besser!"
Entscheide! denn nur
Du hast die Macht,
In sich zu entscheiden,
Uns stehe – zu
Leben bereit.

Again, in 1943 he wrote: "There is no sense in talking about last things, as long as one has no solution for the first. But the first are only to be attained with the greatest difficulty." Glikson by temperament sought conscious rational solutions, however laborious, rather than subjective answers that too quickly glided over difficulties for the sake of immediate satisfactions, however ephemeral. But there was too much of the poet and the lover in him not to have a place for that which can be formulated or expressed only in poetry or music. Purcell's music, in particular the ode to Saint Cecilia, had a special appeal to him. "Though it may sound trange." he wrote me. "I conceive Purcell's music as Architectural and it accompanies me in my work."

To these impressions of Glikson's character as revealed in his own writing, I would like to add the testimony of a colleague, Boas Evron, who worked closely with him and helped him put his sometimes Germanic turns of speech into a more idiomatic kind of English. "This work," observes Evron, was "the most wonderful exercise in intellectual discipline I ever had. Arthur's loyalty to his ideas, his struggle to give them adequate and precise expression, was a lesson in unrelenting intellectual *morality*."

This combination of emotional sensitiveness and intellectual rigor impressed everyone who knew Glikson. Evron's explanation of it, as deriving in its inner tensions from the social shocks of war, and the political and economic chaos that accompanied his entire life, from 1911 to 1967, is the key of course to that whole generation; but not everyone reacted to it in the fashion that Glikson did. "The confrontation with Nazi irrationality," Evron goes on to say, "may have been the origin of his emphasis on rationality, on *planning*, and the conviction that processes must not be permitted to run out their "immanent" courses but be brought under comprehensive, life-enhancing human control."

These articles and lectures opened with a report, on Regional and National Planning in the Netherlands, written after spending four months at the Institute for Social Studies at The Hague: too specific and detailed to be included in this book. But he was conscious of the need for training a new generation of planners who would think habitually in terms of organic complexity, and who would realize that man himself, in order to survive would have, as he wrote me in 1952, to re-insert himself into "nature's life cycles and into a greater scale of time" – a tendency "decisively opposed to disintegration as everyday events reveal it." He would have appreciated the story that another ecologist-planner, Ian McHarg tells, about being chosen, with Lou Kahn, to find a 250 acre site for a Temple of Science. By the time McHarg had picked out the site, Kahn had already designed the building without the faintest concern for its environmental relationship. Glikson knew that this kind of solo estheticism was a formula for hastening further disintegration.

Since the editing of this book has been a labor of love, and since my appreciation of Glikson's contributions is based on a friend's access to his mind, both by letters and by infrequent but intense, indeed intellectually exhaustive meetings, I should perhaps say a word about our relations; for I cannot pretend to give an entirely objective view of his work, if "objective" means "neutral" or "indifferent."

Glikson and I first met in New York, he being introduced by a letter from the Director of his Planning Division, Arjen Sharon; and as people used to say about lovers, we became friends at first sight, for the way had been prepared by common bonds of interest and outlook. In my youth, while studying biology, I had come under the influence of Patrick Geddes who brought into sociology and civic design the fundamental ecological concepts that Darwin, and Geddes's own teacher, Huxley, had developed. Huxley's study of the physiolography of the Thames Basin was an early and masterly example of unified thinking about the entire environment.

In Glikson's development the same role had in turn been played by Ernst Fuhrmann, a generation after Geddes: one of those many German thinkers from Johannes Müller onward, who had resisted the purely mechanistic interpretation of nature, and had remained aware of the far more complex and interdependent realities of the organic world. Geddes, with his gift for systematic thinking, enlarged and gave sharpness to the more intuitive approach of Fuhrmann; and the fact that Geddes had likewise been a practicing planner only made his philosophy closer to Glikson's needs, as the reader of this book will discover.

Since Glikson and I met and exchanged ideas at widely spaced intervals, we had too much to say about our immediate concerns to explore with any fullness each other's earlier history: so about many personal details of my friend's life, I must depend on second hand evidence, or admit ignorance. Our best opportunity to get more closely acquainted came at the eight day Wenner-Gren Conference at Princeton in 1955: the meeting that gave rise to the massive volume on "Man's Role in Changing the Face of the Earth," dedicated by Carl Sauer to the American pioneer in environmental studies, George Perkins Marsh. I remember particularly a wonderful hour, in the deepening summer twilight, walking on the golf course beside our hotel, stimulated and delighted by each other's ideas, and by our sense of a common purpose. Though we seized every possible occasion to meet – in New York, in Amsterdam, and more than once in London and at my country home at Amenia, we never even began to fill up the vacant spaces in our knowledge of each other's life and work.

Through me, happily, Glikson got to know not only the work of Patrick Geddes; but he also became acquainted with the surviving members of a pioneer group known as the Regional Planning Association of America, with Clarence Stein, the dean of American community planners, and with Benton MacKaye, famous as the founder of the Appalachian Trail, but perhaps even more important for his path-breaking work, The New Exploration. But even here lack of time limited the range of our discussions. I cannot, for instance, recall discussing Ebenezer Howard's concept of the Garden City and his equally important suggestion for "social cities," which our regionalist group developed further. Ant I have no notion of what Glikson thought about the New Towns of England, or whether he had actually inspected any on his first visit to England in 1953. Since 1949 some thirty new towns have been built in Israel: so in practice this was a subject in which Glikson was deeply immersed. But because of the stringent need for making the most of Israel's limited agricultural area and water-poor soil, it was only natural that he should have turned to the Dutch reclamation of the polders, rather than to the English need to combat metropolitan congestion and suburban sprawl, for less ons in this new type of urban-rural design.

Though these diverse exponents of a new regionalism, a new type of urban integration, all had something to teach Glikson, they re-enforced, rather than supplied, the ideas he himself had already developed; for Glikson accepted nothing from others without close critical inspection and revision in the light of his own experience and his sense of opening possibilities. Never was a planner less given to accepting current practices without critical examination, no matter how fashionable or profitable they might be. Yet I have known no one of his generation more open to fresh ideas from new sources, more ready to accept what was still life-worthy in old traditions, or more willing to re-establish human values that had been cast aside for the sake of profit-making but often deleterious forms of mechanical progress.

Characteristically, he found the CIAM approach to housing and city building too mechanistic, too arbitrary, too much influenced by spectacular showmanship.

A propos the outmoded methods of the one-time "avant-garde" of the twenties, Glikson observed: "As a leading idea, Geddes's view of the three-fold inter-relations of folk, work, and place seems to me immeasurably more fruitful than the rather technological C.I.A.M. definition of "Living, Working, Recreation, Circulation" as the keys to town planning, or Le Corbusier's "Twenty-four hour solar day . . . the measure of our Town Planning adventures." He fully agreed with my own criticism that a definition of the city or city planning, which left out the central core of the historic city – its religious, political, and educational institutions – was in fact ignoring the central functions and purposes of city life or at best acknowledging their existence in the vaguest possible terms, as "living."

Though Glikson's working base, after 1935, was in Israel, he was neither an orthodox Jew nor a Maccabean nationalist: quite the contrary. Though we never discussed Arab-Israeli relations in detail, I know he was disturbed by those on both sides who took a self-righteously intransigent position. As in Theodorf Herzl's neglected Zionist utopia, Altneuland, he felt that the success of the whole venture in Jewish resettlement ultimately, rested not on a racially segregated state, but upon a cooperative federal union of the Jewish and Arab communities. Characteristically, he went out of his way, at the Wenner-Gren conference, to voice approval of Soliman Huzayyim, the brilliant Egyptian geologist, whose personal presence often vibrated through our meetings. Glikson was a man of the world in the reasonable, eighteenth century sense that has now, in this age of unreason, been forgotten, or dismissed as a utopian impossibility. Human welfare, not just the prosperity of his own tribe or ideology, was what mattered to him; and accordingly he was at home everywhere. The fact that he was emotionally in tune with such different landscapes as those of the Dutch polders, the orchards and vineyards of Tuscany, and the mountains of Crete says much for his open personality, sensitive to genuine life-values wherever he found them. This trait is fully revealed by his marvelous photographs of anonymous buildings and bridges, of peasant and fishermen, in Italy and Greece: the sturdy reminders of a viable way of life and a still living past.

Characteristically, too, Glikson had a passion for trees and delighted to photograph them. One of the earliest photos he sent me as a New Year's greeting was that of a tree, with the following explanation. "The enclosed picture I took in Holland of a tree in a Nature Reservation . . . I see this picture not only as a symbol, but also as the thing itself, a manifestation of natural power. It resembles lightning, but one directed from earth upwards and this means a constructive process, not a destructive one. It embraces and covers space with its spiral shaped rising branches; it creates space habitable for a whole community of organisms – moss, undergrowth, insects, birds, and many more species, for which it becomes a living house. At the same time it is a picture of a dance-like ecstatic movement, again not a symbolic dance, but living matter expressing itself in dancing in spite of the rules of heaviness and destruction. The tree stands on a moor; this means it is a tree which will not grow old because of the unsafe ground, which lets it fall after reaching a certain height. All the more vehement is its will to life . . . To me it seems a most reliable tree in a landscape of mostly soil exploitation and sometimes decorative gardens of illusions."

This description gives both the quality of Glikson's mind, the ambiance of his more systematic and constructive thinking, and the tone of our correspondence. All that is lacking are his sad, humorous eyes and his sensitive smile. Coming on a visit to my own region, the parklike landscape of Dutchess Country, a matrix of meadows, hillside pastures, and wooded hills with their second or third growth forests, or viewing my home-acre, with its towering maples, more than a century old, and the noble elms, one American, one English that have survived the elm blight, he was in ecstasy. Though Glikson had learned to make the best of the dusty desert air of Palestine, he was thirsty for more juicy vegetation and drank it in.

While I am combing through his letters, let me round out the picture of his mind with a quite different kind of observation. Commenting on "The Conduct of Life," he said. "You write on Page 135: "In a world governed wholly by chance only order would astonish, which seems to me right. I tried to change your sentence to say: "In a world governed wholly by purpose, only human consciousness would astonish, which seems to me equally right. I want to emphasize by that the strange role played by consciousness in life and evolution. One has to come to conscious decisions in respect to oneself, neighbors, society, nature, as if neither order nor chaos were pre-established. But consciously wished purposes are never directly realizable. The duality of existing order and conscious purposes (or wish) form the essential reality of conscious decisions . . . I wonder if this has not much similarity to propagation processes: the child is not what the father intends him to be, but his parents' *real* self-expression. And further decisions are steps borne into the unknown, just as a new-born child will never lead a life as predesigned by his parents."

It was thoughts like this that played over and around his practical daily tasks as planner-ecologist and architect. No wonder so many other things were passed over in our correspondence!

Though in discussing Glikson's contributions I necessarily emphasize those ecological aspects where his ideas were more penetrating and profound than those of most of his contemporaries, I do not wish to underestimate his work as a contemporary architect, seeking to realize by legitimate means the highest possible esthetic aims – even under the most rigorous practical limitations. Repeatedly, when we met, he would place his new plans and elevations before me, for critical appreciation: particularly those for the experimental habitation unit in the New Town of Kiryath Gat. No single fresh invention, no technical trick, no exuberance of detail characterized this new neighborhood plan. Its first purpose was not to strike the eye but to satisfy a multitude of carefully weighed and apportioned human needs: the esthetic effect was a necessary part of the whole experience, but only a part. Such a work is not to be judged by photographs – if indeed any genuine work of architecture is ever to be judged by photographs. Nor can it be judged by its immediate effect on the eye before the whole site has been planted, furnished, inhabited, and remodelled by use. But I know Glikson was upset because some of the "younger" European architects whose work he valued did not respond critically to his presentation of the Habitational Unit. All too quickly they had turned aside to consider more striking programmatic projects that were based more heavily on the immediate satisfaction of the architect's ego.

Not that Glikson was indifferent to esthetic experiment. Like Werner Moser, another architect whose respect for constructional honesty and rational order were not unlike Glikson's, he felt impelled in a letter written in 1957 to "say a word about Le Corbusier's Chapel at Ronchamp. Quite contrary to my impression from photos and plans: an astounding thing, most ingenious as a sculpture, as interior space, *and* in the transition from outside to inside. The walk around the building and into it is an adventure in the good sense, with sudden surprises, contrasts which keep you in tension. The symbolism of the whole is extremely sophisticated, and makes you laugh – here is the weakness of the whole thing. It is after all only a good joke, but a *very* good one – this is an important positive indication full of gaiety. There is an incredible amount of fuss and tourism around it which fortunately overshadows any reasonable "religious function" of the

Chapel. The place is overcrowded. But it is truly rich in ideas and forms –
which can never be reproduced in pictures. It is another thing that the "educational" impact of the chapel on the architect-imitators is and will be horri-
ble." It was with these appreciative yet critical thoughts in mind that he
asked earlier for news about the publication of the works of the brilliant
Pole, Matthew Nowicki, the most promising of the younger architects in
America. Nowicki had been killed in a plane accident in 1950 at the age
of 40; but as collaborator with Albert Mayer in the first plans for Chandi-
garh, his designs for housing and public buildings there promised far more
than Le Corbusier's to reunite as never before the functional, esthetic, social, and symbolic aspects of modern design.

What remains now, before summarizing Artur Glikson's fresh contributions to planning thought, is to say a few words about the rest of his all-too-
brief career. Though he gave much of his time to teaching others, mainly in
the Haifa Technion, a large part of his work, whether in travel, in surveys,
or in conference with others, was devoted to teaching himself. And it was
precisely this capacity to receive new ideas and to revamp old ones in the
light of new experience that Glikson excelled. This unceasing capacity to
learn, this willingness to re-examine current dogmas, were what made his
mind so interesting, and his planning projects so full of further promise – and
what made his early death so tragic. In the interests of his own education, in
1952, he had it in mind, he wrote me, "to leave the country for a few years
and to work as a planner in India or else on this side of East Asia. It attracts
me not just as an adventure, but as the most essential experience of the East-
West problem." The reconciliation of Eastern and Western modes of life
seemed to him one of the most essential tasks of our time, unified as our
world is by mechanical and electrical instruments and superficial human
contacts, but still deeply separated in consciousness, and even more in the
unconscious, than ever before. He realized that neither the old imperialism
nor the neo-colonial isolationism was an answer to this problem.
   As if in response to Glikson's private whishes there came what proved to
be his final opportunity, that of heading a team of planners to work out a
Regional Plan for Crete. Such a plan would be broad enough to guide its
further development, not merely expanding its possibility as a productive
living community of villages and farms, but as a place of geographic and
archaeological interest for outside visitors, without permitting all its best
features to be obliterated to increase the profits for the airlines and international hotel syndicates that have created the new factitious industry of mass
tourism. This difficult problem was exactly to Glikson's taste: it brought

forth his well-organized knowledge of places, peoples, industries, vocations, traditions, customs in all their dynamic ecological complexity. To preserve whatever was vital and active in local traditions, while opening up the community to fresh stimulus, and even making possible fresh earning power through tourism itself – this was an opportunity to test his own philosophy in practice. And he even drew upon Benton MacKaye's Appalachian Trail for a Cretan trail that would serve as a backbone of primeval wilderness for the whole island.

The Greeks and Cretans who worked with him on this plan have testified no less than his Israeli colleagues to Glikson's personal qualities in collaboration. He was no prima donna; for though he would fight stubbornly to maintain respect for his guiding ecological principles, he considered himself, as head of a team, primus inter pares; and for all his zeal and firmness, he was an easy man to get on with. Before Glikson died, his plan for the regional development of Crete had reached a definitive stage, which he embodied in a detailed memorandum, accompanied by charts and maps. This report is a model of concise but comprehensive exposition, dealing as it does with both the broad strategy of development over a ten year period and with the practical details for carrying it out. Since it is for professional use, rather than general reading, I have included it, minus the charts and maps, in an Appendix.

Under more favorable circumstances than this period of social and political disruption offers, one could imagine that this ecological demonstration in Crete, intelligently carried out, might have been the forerunner of a series of such regional surveys and plans, carrying further those begun in the Netherlands, in Israel, in the Rhone Valley, in the Tennessee Valley, or in the New Towns of Britain. In such widely scattered experiments, a new order of planning would lay the foundations for the ultimate emergence of a life-oriented economy. But that favorable beginning, like so many other equally promising initiatives in our time, was aborted, not just by Glikson's death, but by the Putsch of the Greek colonels, just as the work that Glikson and his colleagues had begun in Israel is still threatened by the paranoid minorities, Arab and Zionist, whose private fantasies and irrational public aims are not only irreconcilable but suicidal. All of Glikson's life had been dogged by such irrationalities, beginning with the 1914 war and rising to a shrieking climax in the Nazi triumph which drove him to seek refuge in Israel; and similar disorders had invaded his domestic affairs, too. Though he was extremely reticent about his private life, one knew from an occasional cry of anguish that all too often he was deprived of the peace of mind he needed for his work. "Life," he wrote me in 1959, "is grinding me down." That he per-

severed in the face of these collective and individual ordeals gives an indication of his heroic moral qualities – though he would have disclaimed my
use of this adjective.

If Glikson never lacked the fortitude to face the most dismal realities, it
was perhaps because he was quick to recognize and embrace the smallest
evidences of beauty or love which helped to counterbalance, for a blessed
moment, the worst misfortune. One such moment came to him, I am happy
to report, in the years immediately before his death; and his photographs of
landscape and buildings shyly bear witness to it. But before Glikson's work
in Crete came to fruition, his end came, suddenly and unexpectedly, from a
cerebral hemorrhage. Though death tends at first to unify and even magnify
the work of a man's life, well before this happened his associates and friends
realized than he was a personality of the highest potentialities – no mere
professional adept – and that he brought to architecture those deeply human
qualities, of mutual aid and mutual understanding untouched by exhibitio-
nisms, which the masters of modern architecture, with a few notable exceptions, had lacked.

My purpose in this Introduction has been to reveal the various factors
of personal experience, intellectual formulation, and practical effort that
made Glikson's work, not merely important in itself, but even more for what
it may contribute to the future. It wil turn out, I think, that Glikson's vision,
like that of another equally unassuming innovator, Ebenezer Howard, will
prove to be far more essential to the renovation and further development of
the modern world than the grandiose schemes for creating vast megalopolitan
complexes (Le Corbusier, Doxiadis), monstrous megastructures (Kenzo
Tange, Frank Lloyd Wright, Buckminster Fuller), and exorbitant transportation networks that turn the existing cities into a desert of jetports and
jumbled expressways, while they reduce the cultural functions of the city
to those that can be programmed on a computer or brought by television
under remote control. Unlike these leaders, Glikson realized that these
schemes were, in terms of viable biological and social criteria, obsolete.

In negative terms, the human need for the kind of ecological planning that
Glikson stood for has already been fully made out. Except for fanatical tech-
nocrats, who cannot conceive of any alternatives to their own sterile ideology,
everyone realizes that the power system that has dominated the world during
the last century is everywhere breaking down and has already, in many areas,
brought us to the brink of disaster through its massive production of wastes
and poisons, and its rabid commitment to mass extermination as a mean
of maintaining political power. Within a generation, we now realize, the
dominant power complex has done more violent damage to the planet, and

therefore to the further potentialities of human development, than the human race has done hitherto during its entire existence.

Our civilization is within sight of a total breakdown, of which the concurrent evidence of mental breakdown is both a collective symptom and a further cause. This situation calls for an approach to our problems that puts the renewal of life itself in all its multitudinous variety, its protection, its encouragement, its nourishment and enhancement, as the central problem of our age: the one that demands our prompt and urgent attention. And the first need of the planner is not to get more information about this situation, by loading a computer with an even more intricate set of factors, but to take the responsibility to bring into action the sufficient knowledge we already possess, and to apply this knowledge to curtailing, to inhibiting and halting the forces that threaten mankind's existence. To recover the art of being human is the major task of our generation. In order to understand what is meant by "being human" I have dwelt, not only on Glikson's planning ideas, but on his own personality and humanity.

The opening paper on Man's Relationship to his Environment sets the theme of the whole book; while the final essay, published in Le Carré Bleu after Glikson's death, demonstrates the experimental application of his thinking to the concrete planning of an actual habitat. Though Kiryath Gat was only a beginning, it demonstrated how well Glikson had interpreted and been able to satisfy the needs of its inhabitants. Glikson expressed his own pleasure in the already visible results. "It gives me a little comfort to be told that my habitational unit in Kiryath Gat has become the most attractive residential area in the town, while – not though! – it is a mixed unit with all population elements represented; and that in the present building crisis people demand to continue building for them in this unit . . . rather than in other places." A statistical survey made by Dr Judith Shuval in 1965 showed that there was reason for his satisfaction: social cooperation between culturally alien Israeli and Oriental Jews proved more frequent and friendly than in the control unit examined at the same time.

Apart from the larger issues they raise, the virtue of these papers, and indeed of Glikson's whole mode of thinking and working, is that they take equal account of the larger ideological and political efforts that must be made to renovate our entire world society, and of the little steps, open to immediate local action, that are no less imperative – but only if they point in the direction of a larger, life-oriented goal. Not merely the space scale but the time scale, he pointed out, has changed. The more we are able, through instant communication and swift transportation, to embrace and unify the life of mankind as a whole, the more imperative it is that – and here I quote again from a letter – "we face the difficulty of getting man to think in terms of even a hundred years (let alone eternity) which means to behave morally instead of in terms of 3 - 5 years to come, as practised today. This short term thinking keeps life in a foolish, senseless tension and makes all planning "unrealistic" ("Life is time's fool") whereas thinking in longer terms of time involves the creation of what we conceive as values of life, capable and worthy of survival." "But," he added, "as long as the reuniting ideas have not grown out of us, so long the best we can realize are the 'little steps.' " If a million Gliksons were spread over the planet we might at least make a beginning.

By now, I trust, the reader is impatient to hear Artur Glikson's own words, and follow through his line reasoning and his demonstrations. So let me sum up all that I have attempted to say in words I used shortly after Glikson's death in introducing his final essay. "From Artur Glikson's career one thing clearly emerges. He was primarily concerned, not to display technical virtuosity or superficial esthetic originality, but to provide a setting that would do justice to the complexities of nature and the varied needs of human life. His own personal charm, his smiling sobriety, his steadfast rationality, gave his work a special quality. In an age of self-advertisement, he was modest and reticent; in an age warped by crippled geniuses, he was whole and well-balanced: in an age that submitted to its own dehumanization, he restored the human image and returned to the human center. His premature death was a bitter loss; but it will be redeemed if the coming generation understands the lesson of his life and work."

LEWIS MUMFORD

MAN'S RELATIONSHIP TO HIS ENVIRONMENT

I am going to use the term "human environment" to describe the space which surrounds human movement, work, habitation, rest and interaction – towns and villages, settlement influence areas, the rural and the accessible virgin landscape. This environment is generally defined as a set of biological and physical facts in space, as modified by man.

Nothing would distinguish the principles of man's relationship to environment from those of other species, were it not for the fact of his own evolution. Man, a powerful agent of change in space, himself undergoes change in time. This fact transforms his role in any natural community into a dynamic, untable and often contradictory relationship unique among species: dynamic in space – the give-and-take relationship with the earth and its life – and dynamic in time – the recurrent change, destruction, and renewal of such give-and-take relationships in reference to new situations. All established settlement regions show evidence of the continuous imposition of new environmental patterns on top of the old – of a temporal dimension in settlement structure.

Interaction of human and Environmental Evolution

The elucidation of human evolution would provide a key to understanding the changing shape of human environment. But human evolution itself cannot be understood without reference to the interaction of man and environment. A distinction is generally made between the cultural and biological aspects of human evolution. Whereas biological changes are hereditary, cultural acquisitions, such as knowledge, ideas, tools and the environmental heritage can be misunderstood, forgotten or rejected by new generations. Though the two processes make up the changing human constitution, we can discern in man no direct connexion between biological and cultural evolution. However, through man's environmental relationship, an indirect connexion between them is established: first, through the impact of cultural evolution on the biology of man's environment, effecting both qualitative changes in and distribution of plants, animals, water, air, soil, micro-climate, materials, etc.; next, through the decisive influence of these changes in their turn on the total psycho-physical position and function of man in life.

By observing and interpreting the evolution of man's relationship to environment we gain important clues to the profound significance of cultural evolution for the human constitution. In environmental change, cultural facts are converted into biotic realities, while biological conditions mould human culture in their turn; human and environmental evolution become interacting processes of change, constituting a whole life-system. To comprehend the system, we must study alternately its human and environmental components, consider both biological and cultural aspects of evolution, and note their interconnexions whenever they appear.

Environmental Evolution and Environmental Change

When observing the human environment, we are always confronted with its dual determination. It is conditioned by human life, and its conditions man's functions and development. But the evolution of the human environment appears as a sequence of stations, on each of which a high degree of integration of community and environment is realized and maintained over long stretches of time. Communities may come close to identifying themselves with their environment. In various ancient farming societies, such as the South-American Ayllu and the widespread Asian Savah, community and environment were named by the same term. Such knowledge can help us to understand past environmental orders; it can also give us confidence in the future when such an harmonious man-environment relationship might be re-established.

When we wish to effect change in the man-environment relationship, the realization of its mutuality is a necessary starting point. But the truth than man and environment determine and are determined by each other is not sufficient to provide guidance for man's modifying actions. On the contrary, this knowledge is liable to hide the fact that the incentive to change results from the experience of an incompatibility between the environmental situation and the human situation.

Environmental change is motivated either by the existence of free human energy, not needed for the maintenance of existing communities, or by a disturbance in biological or physical equilibrium. A surplus or a deficiency of population or resources, migrations or wars, may initiate a process of environmental change. Such conditions urge communities to "reflect" themselves in new ways upon the surrounding biota and landscape, to introduce different types of land use and to reorganize their cultural habitat. But, in the course of this activity, they encounter reciprocal environmental influences which modify their goals and way of life.

Environmental disequilibrium stimulates thought and action. Yet no definite programme of action can result from the mere awareness of estrangement from environment. The disturbed relation, therefore, urges man to search for guidance on a different level. So long as such guidance is not found, men have no choice but to justify or permit arbitrarily any environmental effort or neglect as a "natural" or a "human" process.

Guidance for environmental modification and cultivation ensuring a balance was gained in the past by instincts, or by myths or religions which established doctrines about man's situation and responsibilities in the universe. Our understanding of past environmental relationships remains incomplete, if we disregard such spiritual bases. In our time we must seek for such guidance, first of all in what Aldo Leopold called "a land ethic ... reflecting the existence of an ecological conscience" and evolved as "an intellectual as well as emotional process"; this ethic, combined with a social conscience, would derive from a sympathetic attitude to life and culture in its various manifestations, leading man to realize his obligation towards the biological context as well as to humanity. Secondly, environmental modification must be guided by an environmental art – and architecture – which not only integrates experience, knowledge and ethics in creative design and development, but which in the very act of integration and design becomes exploration of environmental realities and values. In thus adopting the function of responsible agents of natural and environmental evolution we may hope in time to fulfil our humanity.

Varieties of the Man-Environment Relationship

Under more or less permanent climatic and geological conditions, and barring unusual biological events, there exists only one ecological climax development in a landscape which is devoid of human influence. But in the history of the humanized landscape, there are many "climax" developments because of the changes in the ecological role fulfilled by man. To the extent that a species is characterized by occupying a particular position and fulfilling a specific function in an ecological community, man, by virtue of his cultural differences and their development in time, represents, as it were, a multitude of different species. In respect to the environment man is not a consistent biological unit preserving its identity, but an organism in the process of change. We must therefore consider a wide range of environment contexts. In the following pages I shall give examples of such contexts, as related to human mobility, sedentariness and urbanity. In these brief descriptions, a community's characteristic manner of using land, space and time is shown to emerge from or to result in a pattern of human rest and movement in the landscape. This pattern modifies nature, forms the humanized environment, and implies the creation of specific values in the man-environmen relationship.

Mobility and Sedentariness

The hunter's and collectors' society must be a starting point for any study of the development of the man-environment relationship. Human communities, like other animals, keep moving from camp to camp to replenish supplies of food and water. In their natural ecological state, most regions can sustain human life for only part of the year. Man's main energy "investment" consists, as a result, of his own movement from one favourable ecological context to another. As these optima change with the seasons, it can be said that man maintains his existence by a wise use of time rather than of land. In this context the landscape is formed by ecological systems in which man is no more than a temporary partner. To leave the land alone and to follow the sun is the rationale of man's mobility. Its scale is almost unlimited by natural regions, bringing man into close contact with a wide range of conditions and situations. Intertribal contact is casual, and the general population density is estimated to be in some regions as low as 0.2 persons per square kilometre. Space is experienced essentially as exterior, a seasonally changing unbounded condition. Just as the flow of biotic energy moves continuously with the seasons through the ecological "chains" of the landscape, human mobility likewise has no proper centre though certain locations may be seasonally revisited. The apparent exception to this order of mobility – early riparian settlement – proves the rule, for here we find an exceptional dynamic environmental condition: the regular movement of a whole ecosystem, the river, combined with a favourable climate enabling man to stay in one place, while enjoying (as a beachcomber) an ever-replenished supply of food, materials and water.

Sedentariness

Sedentariness is perhaps the most astonishing biotechnical "invention" and the most successful revolution ever carried through in man's relation to his environment. The conditions for sedentariness are the acquisition of biological knowledge and the faculty organization of biotic processes. Man creates an artificial landscape, in which fertility and water are preserved by actively fulfilling the rule of return observed in nature. The disturbed cycles of growth are ever-restored by transportation, storage and the use of seed, fertilizers, water, crops and materials. This requires a continual investment of human and material energy to maintain the artifact, implying far-reaching environmental change. By this achievement, man necessarily alienates himself from the ecological context; but he uses his new position to establish a relationship to the land on a new level. On this basis mountains are shaped into cultivated step-pyramids and valleys are converted by irrigation into huge containers and renewers of fertility. The earlier instinctive integration of hunters and collectors in the natural space and time is developed into the active identification of farming communities with their environment. Land becomes a community's own ground, while a community belongs to a specific landscape. Inter-communal contacts are characterized by whether they occur on one's "own" or "foreign" territory. Man assumes a central position both in the eco-system and – physically – on his land. The new order embodies the creation of a concentric environmental system, consisting of a nucleus – the place of human habitation, storage, processing and community life – and of peripheral areas of use of land, water, forest, serving a different frequency of human movement and a different intensity of land use.

Human mobility is now essentially confined within the village lands. It assumes the form of a radial flow of men, commodities and energy from the centre to the periphery and back. Within this space man cuts down most of the natural vegetation and replaces it by species which support the qualitative and quantitative increase of his own kind. But the maintenance of this artificial environment and its defence against the return of natural vegetation call for such constant human attention that man becomes inseparable from the very vegetative process he introduced. His life now bears the characteristic of being both animal-mobile and vegetative-stationary.

By cultivation man shortens the time which a natural recuperation of fertility would require. This intensification makes possible a considerable increase in the size of concentrated communities and of population density in general. But sedentariness also means man's increased vulnerability to seasonal changes of climate, floods and drought. To meet these dangers, and to create a bearable domestic micro-climate throughout the year, permanent houses are built and grouped in village clusters. But the characteristic determinant of environmental structure is the organization of biotic processes.

Urbanity

Town building represents an additional artificial arrangement in the human environment. Since the land can be made to produce more food than is required by the primary producer for his own and his family's needs, it can also supply a "surplus" urban population with other aims than land cultivation. In principle these aims are social: the creation of a new ecological system of social interaction, of classes and professions, in which the farming population is only one layer, though the basic one. The urban population considers itself as no longer living from the land, but by the mutual exchange of goods and services among men. The individual now belongs not to a particular landscape but to a social organization spread over a whole region, which may contain villages, hamlets, forests, water resources, etc., and at least one fortified urban nucleus. This centre absorbs a major part of the region's products, and from it the region is ruled. The human determinants of land use are a sociological order on the one hand and an agricultural order on the other hand. Where these orders exist in mutual harmony, men find their living space extended over a whole region. The radius of human mobility and contacts increases enormously, and the order of rest and movement changes in character. The market appears as a central focus of exchange; land use, even in far-away villages, may become more specialized, or more adapted to the natural character of the land. The urban cross-roads are the main points of regional and inter-regional meetings.

Ancient trends towards a more complex settlement structure and trends towards a greater mobility imparting a greater degree of independence of the land are integrated in the town. The town develops into an organ of country-wide and world-wide material and cultural contacts. A new dimension is added to man's environment. With communications on a world scale, many towns may acquire wider super-regional economic, social and cultural functions. Potentially there are no limits to urban growth. A professionally and culturally diversified town is essential to provide a sound basis for regional coexistence. This fact finds expression in the greatly diversified physical structure of towns, differing in principle from the village structure, which consists of groups of equal cells, the farmsteads.

Space in the rural-urban regional environment is converted partly into interior and partly into exterior space. Arrival and departure become important objects of environmental experience and creation. In order to communicate visually and audibly with the region, urban centres rise in height and towns become three-dimensional regional objects.

The artificial social structure of the region becomes somewhat analogous to the principles of biotic structure, for in both cases stability is safeguarded in the complexity of its composition by characteristic kinds and numbers of population. Equally noteworthy is the fact that in many cases the combination of sedentariness with non-agricultural settlement, both man-made situations, results in a new adaptation of the human environment to a natural regional framework, namely the watershed-basins. But at the same time, life in the new environmental structure becomes a matter of precarious balance. Only a step divides urban-rural mutuality from exploitation, surplus productivity from soil-exhaustion, inter-regional contacts from wars, the function of the town as a co-ordinating and distributing organ from that of a parasite.

Evolution of Environmental Values

The ecological integration in the primordial landscape and the time use of the early hunters and collectors form a human trend which can never be lost. It reappears in most sedentary cultures, in the desire for contact with new ecological life-systems, in landscape exploration and later in the ideals of "a return to nature," and in the recreational aims of recovering the sensation of life and mobility in undisturbed nature.

Sedentariness is a cultural achievement, not because it ends the more animal and mobile link of life and replaces it by another, but because it succeeds in uniting the animal and vegetative trends of environmental relationship in a new spatial order of rest and movement, and because it combines land use with time use in a new pattern of human life. One of the contributions of sedentariness to the quality of man's relationship to environment is that man can identify himself with an environment by feeling that he belongs to it, and by being aware of his obligation to maintain it. It is a central motif of the man-environment relationship, which we shall always endeavour to recover in some form in the course of environmental modification.

The town is a superior human environment, because – or rather, if – it constitutes a place and organ of inter-communal unity and cultural continuity. In it, mobility and sedentariness, biological and social systems, poor and rich regions, the various cultures, the individual and the community, the past and the present, all come in contact, coexist and further each other, without losing their identity. This function has been most thoroughly elucidated in Lewis Mumford's *The City in History*. Today, in regional planning, we regard the quality of an integrated urban-rural region as an ideal to be regained in the improved environment of densely populated regions, though we cannot restore it in its past forms.

The sequence of these stages, and of others not mentioned here, constitutes what may justly be called "environmental evolution." The human environment is an organization of relationships and facts, but the evolution of this organization is in essential ways analogous to the processes of biological evolution. This evolution may be understood as the continual grafting of new types of environmental relationships on earlier types. It is not a disconnected sequence of incompatible attitudes displacing one another. At each stage specific values have been created and, in the sequence of stages, an accumulation and integration of these values has been accomplished. In this way, the reality of the man-environment relationship has been progressively extended and intensified, and so has the scope of environmental problems to be solved by man. Evolutionary continuity creates multiplicity, and integration becomes a recurring task for each new generation.

How has this "grafting" of the values of environmental relationship been accomplished? Carl Sauer gives a plausible explanation of the processes of cultural and environmental transformation. According to him, the Palaeolithic, Mesolithic and Neolithic represent a sequence of cultural stages through the integration of diversified skills and ideas originating in different regions. He considers that hunting, plant cultivation and the keeping of domestic animals were introduced to Europe, for instance, by outside people. Most cultural trends have specific "hearths" or points of origin. Environmental evolution and the related human evolution have been connected from the beginning with world-wide human interaction and cross-fertilization of diversified ideas and values. The process consists largely of the assimilation and integration of "transported" practices, attitudes and values into existing relationships, with the result that a new type of environmental relationship and structure is formed. Today, with the dispersal of technology and civilization from two or three world centres, we observe such processes in all parts of the world, but they are of ancient origin.

I have offered a few representative illustrations of the growth of man's multiple relationships to his environment and of the progressive increase of the complexity of the environment as a fact in human life. They suffice to imply three fundamental values which must be aimed at in any attempt at the creative modification of the environment, and which have been valid since the first city was founded.

1. The establishment of life-enriching contacts with places and communities through an order of mobility.

2. The self-identification of man and community with a modified space and biota, to be achieved by a conscious relationship to environment, which can be realized only in an order of sedentariness.

3. The co-ordination and mutual adjustment of human and material opposites and diversified ways of living in composite environmental structures, in towns and urban regions.

Environment of Metropolitan Civilization

In view of these values, examination of contemporary trends reveals the extent of the present human and environmental emergency. Our answer to the problem of increased land requirements for habitation, production, circulation and recreation has been the specialization of land use, mainly in the highly developed countries. But, while the specialization of land use for crafts, trades, crops, etc., on the small scale of medieval and renaissance towns became an asset to environmental utility and beauty, a specialization on the huge scale of contemporary population concentrations disrupts the relationships of a community to the regional environment. The landscape has been subdivided into mutually exclusive or conflicting land use areas for the production of maximal amounts of food, timber, water and energy to supply the increased concentrations of population and the increased requirements of a high technological standard. The necessity to introduce new land-use types and methods has led to the implementation of powerful and quickly superseded stopgap measures provided by science and technology. The new type of land use has not led to a new relationship, but only to an increasing alienation of man from his environment.

To connect specialized areas of land use and production with marketing and transportation centres and with the comparatively small number of great urban population concentrations, the net of country and worldwide communications has spread enormously. But the contacts thus established with communities and landscape have led to crude colonial and commercial exploitation of population and land. Economic inequalities between industrialized and colonial or developing regions have increased and must increase further, as long as these world-wide relations are controlled merely by the free play of market forces. Technology and international communications have spread only the meanest cultural achievements uniformly over the whole world. Equally, the increase of mobility has not led to increased contact with and access to landscape and towns. So far, the space crossed by transportation is not considered and treated as a new field of human experience, but as a vacuum or an obstacle to be bridged by streamlined channels of mechanized movement, planned with the single aim of connecting several focal points of interest.

Some metropolitan centres have become, to an unprecedented degree, the meeting grounds of populations, products, cultures, knowledge and materials from all over the world. In principle this meeting of opposites might breed a new cultural integration and an enrichment of life. In the contemporary metropolis it leads to fighting among groups and individuals to the extinction of less resistant elements. It does not necessarily appear as physical murder; but increasing uniformity and mechanization bring about the same result in respect to the quality of life, the annihilation of social and environmental complexity. Uniform housing and town extensions designed for the fictitious "average" family and citizen turn urban conglomerations into "anti-town." They contradict the *raison d'être* of urban life, the co-existence and co-relation of diversified elements.

Fortunately, however, metropolitan mechanization is far from achieving its end, and we have the more hopeful condition of metropolitan chaos. The almost complete confusion and interference of all kinds of movement with all kinds of rest in the central areas contrasts with the urban periphery where the growth of suburbia represents an attempt to escape from both the problems and the idea of the city. The enormous confluence of human life and interest in the metropolitan city has resulted in self-destructive building densities, and haphazard overlapping and mutual hampering of areas of habitation, industry, business, storage, recreation and education. The private car has expelled civic life from its natural centres and from many residential and recreational areas. Unsolved traffic problems are only a physical expression of the confusion ruling all human communication and interaction in the metropolis and even in smaller towns.

As the urban environment becomes a dangerous nuisance, the same civilization provides the individual with protection from direct social and environmental contact by means of telecommunication, drugs, and the provision of an illusory environment through illustrated papers, the cinema and television. Technology is thus used to prevent, to sterilize or to serve as substitute for direct contact. Technological inventiveness in itself is not the disturbing factor in metropolitan life; the mischief comes from its use, first in misraising environmental fears, then in creating "protective" measures which wrech vital relationships of man to environment. Finally the effort of the metropolitan population to find refuge in the natural or historic environment is being progressively cut off by the commercialization of recreation or by the recreational mass-movement itself, defeating its own purpose. Only the richest classes – and outsiders – can afford to get away far enough from the great concentrations of population, by motor car or jet plane – and then only so long as their number is small.

Huge quantities of human work, materials energy from all over the world are invested to maintain life in the highly developed urban centres of the world. This environmental over-development must be put in the context of the under-development of urban agglomerations in Asia, Africa and South America; there the economic surplus and technological stop-gap measures are not available, and therefore an even graver crisis of environment in-compatibilities has developed. The number of people living under intole-rable conditions of poverty and overcrowding in these few overgrown shanty-towns is steadily increasing with the influx of new multitudes seeking survival. In our "One-World," these extreme contrasts of wasted wealth versus wasted humanity cannot exist concurrently for long, be it for politi-cal or moral reasons. Because of its functional inner-incompatibilities and the absundity of its relation to world problems, we must assume that the metropolis in its present form will prove to be a short-lived phe-nomenon. Its breakdown would have decisive effects on the vast-regions which depend in some form on metropolitan centres. The future of big cities and their regions of influence has therefore become a problem of critical urgency. The relationship of the metropolitan inhabitants to their environ-ment has taken on an "explosive" character. A revolutionary attitude is essential.

In vast areas of the world, man has become a pathogen, a disease of nature, and there is a high degree of probability that, as Marston Bates says, "when the host dies . . . so does the pathogen."

Aims of Environmental Renewal

It is an open question whether our period will be able to add anything new to the general aims of environmental relationship; but their recovery and adaptation would open up an entirely new and hopeful prospect for environ-mental renewal and human evolution. Renewal involves a readiness to res-pect and be formed by external biological and physical conditions. As breath-ing is vital to the maintenance of life, so an alternating passive-active rela-tionship seems essential for human and environmental evolution.

In striving for such participation, our starting point should be awareness in particular places, in relation to the "One-World" scale of the contem-porary human situation. Of special relevance are population increase, me-chanized mass production of commodities, enhanced mobility, the meeting of divergent cultures, and the world-wide interrelation of human and en-vironmental problems. Under these conditions, the achievement of the fun-damental aims of the man-environment relationship constitutes at once an enormous difficulty and a unique human opportunity.

It has been argued that such conditions determine the future of man and environment. But I believe that they are entirely "neutral" facts, and that the quality and direction of human action will determine whether, and to what degree, they will effect even greater environmental disaster or a renewal and enhancement of "environmental breathing." World-wide human contacts are potentially a means to individual and communal enrichment of life; we have reached a state in wrich the old ideal of the unity of mankind can be made a reality. This unity should lead to the introduction of new influences and forms in environmental creation, even on the local level of planning new towns and villages, matched with the preservation of individual and regional identity. Population increase offers a chance to intensify land use and settlement, to build better towns and villages with the purpose of enhancing the identification of man with the environment. Mass mobility makes it possible to belong to a particular place yet to enter into direct contacts with a world-wide environment and with other populations. The technology of telecommunication and transportation can be applied to the desirable location and distribution of settlements. The mechanization of production and construction should serve to meet the vastly increased requirements of population and development, and the demand to share the achievements of the world community. The meeting of cultures, economies, generations, and ideologies in all developing regions, calls for their mutual adjustment, or integration in a new composite environment.

The world-wide interrelation of environments is a new fact which has emerged in our own times. By accepting this situation, we might arrive at the formulation of a new aim for the man-environment relationship, namely continuity – meaning the world-wide spatial continuity of the humanized and natural environment as well as its temporal continuity integrating the natural and historic environment with contemporary creations.

Design for Action

With our technological equipment and surplus economies, we are in a better position to tackle large-scale problems of amelioration, transportation and construction than any previous generation. But our comprehension of the biological and cultural meaning of the changes occuring in our immediate surroundings has weakened.

Experience has shown that we cannot trust the deterioration of environmental conditions in over-developed metropolitan or in poor over-populated regions to lead dialectically to a re-orientation of communities. If environmental deficiencies and an hunger for a more "natural" life drive man into revolt, the result is more likely to be the total destruction of populations and culture than a search from the ground up for renewed vital contacts.

Our best chance, at the moment, seems to lie in the creation of "seeds" of future action. This is the first function of environmental planning. Designs for improvement must be ready at the moment of crisis, or, preferably, before it is reached. Environmental design is not a stop-gap measure, but the initial stage of a conscious evolutionary process. The means and ends of environmental change must be mentally and experimentally prepared by small nuclei of architects, biologists, philosophers and sociologists who have become aware of the full extent of crisis.

Environmental modification must aim at the intensification of life, both by the strengthening of its roots through better functional arrangements, and by the elevation of the man-environment relationship to the level of a psychic experience. This integration of functional and spiritual aspects of environmental structures in a rhythmic order is the subject of art – the architectural design of buildings, settlements and regions. Architecture, thus conceived, would result from "the passage of the world into the soul of man, to suffer there a change and reappear a new and higher fact" (Emerson). Architecture creates a new level of psycho-physical relationship of man to surrounding life and space.

On such a basis, an experimental architecture might start with design for specific people in a specific environment. In the process of design, the images of environmental renewal should become suggestive of orders of rest and movement, of human "breathing" in the environment, of passages from exterior to interior space in landscapes, cities, quarters and houses. The environment should speak a "language" of contact with natural facts, social interaction and a sensible interrelatedness of the natural and the artificial components of human environment. Experimental architecture would have the pioneer function of forming both contemporary architect and architecture, of accumulating experience and of inspiring people to interest and participation in environmental change.

Beginnings of Environmental Renewal

The conviction has grown that the scale and scope of contemporary environmental needs renders reliance on trial-and-error methods of development both immoral and dangerous, and that comprehensive planning for environmental reconstruction has become imperative. The first and greatest exponent of this movement was Sir Patrick Geddes. As a biologist, geologist, sociologist, geographer and planner he conceived the full human importance of environmental renewal and of the intensification of the man-environment relationship. In spite of great building activity and a vast increase in public and private planning organizations, the realization of these ideas in our own "second industrial revolution" stands in no proportion yet to the growing emergency, the technically and economically improved conditions for large-scale development operations and the knowledge gained in planning procedure.

So much "development" is going on that it seems important to point out those beginnings which are distinctly oriented to comprehensive human-environmental renewal. First, there is progress in the planning of new urban quarters. In a number of countries the earlier monstrous uniformity of public housing is being gradually supplanted by the dual priciple of "variety in unity and unity in variety," both in the social housing programme and in the actual urban building forms. On the basis of a positive attitude towards the urban quality of living, a spatial framework for coexistence and relatedness of various human elements, has often been successfully established, at least on the level of the urban sub-unit. It is however, significant that our time has not yet produced an idea of the contemporary large city.

Of particular importance are the few realized schemes of regional planning and the influence of its ideas on other facets of planning. Regional planning is often erroneously interpreted as an attempt at artificial isolation of region-wide populations from international cultural, social and economic influences. But this is contrary to its basic conceptions and intentions. Lewis Mumford has emphasized the essential unity of regionalism and universalism. In his conception, man's relationship to environment should extend simultaneously on various environmental levels, such as the small community, the village, the town, the region, the country and the world. If one of these links is missing, the interaction between the individual and the larger communities is invalidated, and man's relationship to environment is degraded to isolation or disruption.

So far as regional planning is concerned with the renewal of a specific area, it aims at the re-establishment of human self-identification with the environment; but a regional framework of settlement, work and recreation is inherently related both to the smaller framework of local communities and to the higher level of super-regions or countries. Regional planning therefore is also a means of establishing a graduated relationship between the individual and the world.

The few existing instances of regional schemes, as in the Tennessee Valley Authority, Holland and Israel, exemplify both a trend of universal regionalism and of regional universalism. It also seems important that on the basis of this limited experience, proposals for the reconstruction of the regional landscape as the environment for both settled and mobile, rural and urban, working and resting populations have been adopted in principle by several international organizations.

As a recent experimental step towards environmental improvement, I would like to cite a single building, an orphanage, built two years ago in Amsterdam by Aldo van Eyck. The design of this house is a result of the intellectual, ethical and artistic comprehension of the quality of movement and rest of specific people in a specific environment. The conceptual basis of the layout is, in the words of the architect, "... the reconciliation of opposites," such as the life of the individual and the collective, the experience of interior and exterior space, of lanes of movement and interaction, diversity and unity, light and shadow. These opposites are transformed into "dual-phenomena" which cannot "... be split into incompatible polarities without the halves forfeiting whatever they stand for." In the building the transitions from one particular space to another are articulated by means of "... in-between places ... providing the common ground where conflicting polarities can again become dual phenomena." Achievements of modern building technology are applied in this house to serve its function and the development of its human qualities rather than to exhibit technology. In fact, this building, housing 125 orphans, represents a little city within a city (Amsterdam), and it is both a realization of a unique composite architectural idea and — surprisingly — of a contemporary conception of a city in a nutshell.

Human Importance of Environmental Renewal

Environmental renewal has a central function in the improvement of the material and spiritual condition of man. In a world of expanding physical communications, orientation on environment would present a way to a new integration, as summarized in the following points:

The environmental problem embraces the basic level of human amenity and the highest level of human evolution, its biological and cultural aspects, the individual and the collective.

The human relationship to environment is a way of establishing interaction between human and extra-human life. This interaction is stimulated by the very expression of man's new position and function in the biosphere.

The awareness of the environment, as well as the creation of new and meaningful environmental structures, would engage and relate direct experience to science, to ethics and to art. Environmental renewal may thus become a way to fuse the separate branches of culture into a single, composite structure.

The environmental problem is in character, and today even in extent, a world-wide problem of human interaction; environmental renewal, therefore, is a means of approaching the unity of mankind on both the material and the spiritual levels.

The realization of environmental relationship is a matter of carrying the values created in past contexts into the present; environmental renewal might therefore be based upon the recognition of human continuity, and find expression in the integration of old and new values.

The mutual adjustment of opposites in a new environment, the increased material and spiritual contact among world regions, and the renewed attachment to a particular environment, might constitute the positive contents of peace.

No layout of the future world-city or of Utopia is presented here. Often the image of a progressive future is only an attempt to escape from the most decisive problems of the present. Environmental renewal is a continuous process originating in the matrix of present conditions. I have laid emphasis upon the problems of environmental values and attitudes. Once these values and attitudes are realized, environment, as a projection of an enriched human life, may assume an unforseeable multitude of functions and forms. To effect evolution instead of expecting its advent, we must reintegrate values of the man-environment relationship created in the past, to cross-fertilize contemporary cultures and to relate all of them to the present situation of man and environment.

Our concern with the fate of future generations mus lead us to the intensification of our present lives. Thus man, instead of resigning himself to the tyranny of historical determinism, will impose a pattern of continuity on space and time.

II

RECREATIONAL LAND USE

Origins of the Recreational Movement

A general survey of the origins of the problem of recreational land use reveals the following relevant stages of development of many industrialized countries during the last century or so:

1. Large numbers of peasants and peasants' sons gave up their ancient relationship to the soil and village, leaving their rural environment to concentrate in towns and seek employment in industries and services. Overnight, small urban or rural settlements grew enormously, both in area and in density of habitations, so that huge tracts of the surrounding landscape underwent urbanization. This expansion of urban political and economic power into the countryside, and urban methods of production and commerce, led first to a growing economic utilization of rural resources and later to a gradual deterioration of the rural and indigenous landscape by deforestation, mechanization of agriculture, parcellation, introduction of monocultures, faulty methods of cultivation, mining, and construction of industrial and power plants. Soil erosion, disturbance of the water cycles, and loss of fertility and of beauty of landscape are the symptoms of man-made land disease.

2. The still increasing urban population, compressed in quarters where unhealthy conditions prevailed, remote from the open country, began to sense what it had lost and raised a demand for temporary environmental compensation. The rural and indigenous environment became for the urbanite a recreational environment. The peasant sons still wished to return to the country for a holiday. Gradually the need for recreational facilities to maintain the health and efficiency of the urban population became recognized. However, during the period of urban expansion the original cultural landscape had been largely defaced and turned into the "steppe of culture" – as the Dutch call the new rural pattern. Only isolated parts – often spots of economic decay – had kept their original rural character.

3. Pressure of vacationers on the remaining rural and indigenous places and on newly established resorts became violent. This very pressure destroyed these places as true resources for restful recreation. In the attempt to escape overcrowding and noise and to rediscover landscape, holiday-makers were driven ever farther away from the cities. Gradually, social and medical demands for recreational areas for the inhabitants of big cities became incompatible with the physical limitations of, or distance to, recreational land. The recreational movement of the population was hampered, and, as the crisis became obvious, there originated the problem of recreational land use.

The recreational movement should be considered as belonging to the wider contemporary phenomenon of population movement to and from the big centers – of spatial contraction and expansion of resources and commodities, of people and ideas. The most obvious and well known of these phenomena is the tremendous concentration of population and produce from the most distant regions in metropolitan and other big-city centers. In the dynamics of city life the demand for recreation represents a reaction against the psychophysical complexity of life introduced by centralization and industrialization, and reveals a tendency to reverse the prevailing spatial relations. It is an attempt to balance the centripetal concentration by centrifugal diffusion – by a temporary escape back to the places of natural and historic origin of the people: to the indigenous and rural landscape, the hamlet, the little town bypassed by modern development – in the hope of restoring, of "recreating," health, energy, and mental equilibrium.

We have little evidence of specifically recreational land use and facilities for preindustrial periods, because they represented a wholly integrated and therefore unrecognizable ingredient of environment. Private gardens and orchards, large public squares, the well, the streets, and the near-by surrounding rural landscape, all in the context of but moderate housing density generally, provided for the recreational needs of the medieval citizen. In ancient health resorts, such as Bath in England, Tiberias in Israel, and Epidaurus in Argolis, people were not directly seeking relaxation and change, but rather the healing qualities of air, water, and places. In comparison with our century, any recreational movement of former times was composed of a mere trickle of population, "confined to well-to-do folk and beset with difficulties of communication" (Abercrombie and Matthew, 1949, p. 141).

The appearance of a demand for recreation is evidence of the loss of environmental integrity. When residences become mass dwelling machines and factories become poisoned prisons, the "natural life" becomes an ideal. The ugliness of the places we pass through during daily life stimulates a yearning for purified beauty during a period of rest. "Natural" and "beautiful" become notions attaching to a part-time recreational existence. To compensate for these irritations, a new specialized function becomes a social need of city life, and therefore the destiny of special extra-urban areas of forests, river-banks, mountains, beaches, memorable places, as well as resorts: *recreation*, promising pleasure, play, and adventure, all in a concentrated spatial and temporal capsule.

Recreation it not, however, confined to outdoor holiday-making, though this is at present its most conspicuous part. Let us for a moment consider recreation as a biological need, an ingredient of the rhythm of life: effort – relaxation, toil – leisure, routine – adventure. It has its place, then, in the life-maintaining functions in the same way as exhaling is necessary to the physical maintenance of life at any moment. The most important means to achieve recreation in this sense is considered to be a change of environment – we are inclined to say *any* change, the more radical the better. Whereas townsmen migrate to the open country and to the seaside, the farmer looks for recreation in the city. As a counterbalance to the daily way of life, people may search for recreation either in solitude or in crowded centers of amuse-ment, either in closed space or in open squares. Because of this variety of individual demand for recreation, we include in any enumeration of re-creational facilities establishments as different as a coffee-house and a park, a swimming pool and a historical site, a pleasure garden and a whole river system with its fishing and boating facilities, a holiday resort and a wildlife reservation.

The motives driving man to search for recreation in change of environment have not been sufficiently clarified. In many cases it is possible to explain recreation as an attempt to return to lost environmental values and ways of life. Among the most desired targets of such recreational return is the pri-mitive social life of hunting and berry-gathering – primitive in food, shelter, clothing, habitat – though other people may content themselves mainly with rediscovering the indigenous environment in solitude.

It is also possible to assume that there exists in man a biological urge to employ his ability to change his environment. This ability, characterizing animal life generally, is even more the achievement of man, who can adapt himself artificially to varying environments; it is especially exhibited by urban man. But he has often little chance to exercise that ability in the daily run of life. The trend to move about reappears, then, as a recreational need. For we find recreation in just what we had to forego in daily life. To come in touch with different types of environment belongs probably in the same category of desires as the physical demand for a variegated nutrition and the psychic demand for variegated social contacts.

It might be possible to see a parallel between the motives behind recreational mobility and those behind nomadism. With the pastoral nomad, it is the low grade of fertility and carrying capacity reached by land after a period of pasturing which compels him to travel in search of unexploited régions. Similarly, a modern urbanite could be considered to be "undernourished" in respect to environment. The recreational movement, therefore, is a proof of the inter-relation between man and his physical environment. We detect the importance of environmental variety as a resource of human life because we miss it, especially in a time like ours, characterized by the low quality of our artificial urban environment.

The need for recreation varies with the individual; it obviously depends on personal versatility as well as on the quality of his daily environment. To consider recreation as a human need in past, present, and future, we shall have to make a clear distinction between the normal demands for change of environment on the part of members of healthy communities and the abnormal recreational insatiability of modern men living compressed in cities which are not planned to the human scale and which time and again compel attempts to escape.

The Recreational Crisis

Land Requirements of Recreation

Recreation by change of environment is a need felt in all the temporal frameworks of life: times during the day, the day itself, the week, the yearly seasons, and lifetimes. Though individual variations are huge, the life of man may be considered to be intersected by periods of recreation (or the desire for such periods) which help to revitalize the cycle of life, to maintain its rhythm by confronting man with change – different environment and food, association with different people or substantial isolation from society, different occupation, and a different feeling of progress of time.

In our civilization each of these types of time periods can be related in a general way to types of spatial frameworks which provide for the needed recreation of man: the family house, which has to serve recreational needs during parts of the day; the public gardens, squares, playgrounds, amusement and cultural centers, which provide for the daily and some of the weekly recreational needs; the city surroundings – with their parks, forests, rivers – where recreation will be sought by many on weekends; and the region in which one's city is situated, in which it should be possible to stroll about during different seasons of the year. Obviously, this series of time-space correlations with types of recreation can be further elaborated; for example, the "migratory periods" of youths and adults, striving to escape any environmental frame or to turn the whole of the earth into their recreational framework.

In town and country considerable tracts of land have already been reserved exclusively for recreational purposes, and ever more are being demanded. For certain countries the amount of space needed per person for recreation can be calculated on an empirical basis. These amounts are much larger than is generally assumed. In an average European home planned for a family of four to five persons (about 85 square meters) at least a third of the built-up area may be considered to serve indoor recreation during parts of the day: leisure within the family circle after a day of work, play, or solitude in reading, writing or mediating. During parts of the year recreation will also be pursued on additional private areas, such as terraces, courtyards, or directly accessible gardens, which in numerous quarters take up 40 - 50 percent of the total land requirement of the neighborhood.

Calculating the land areas needed per family, according to British standards, for public parks, squares, playing fields, and cultural and amusement centers, we again meet the proportion of approximately one-third (about 110 square meters) of the total land requirement of a neighborhood (Abercrombie. 1945. p. 114). The importance of such areas for physical health has often been emphasized. They are also socially essential; besides the bonds formed in an urban society by work and trading, these urban recreational areas are the places where community bonds are formed during leisure time. But in the space allotted to recreation it must be possible as well to find spots for solitude and rest.

Summing up, the land required for such a recreational program within a well-planned neighborhood amounts to more than 70 per cent of its total area. In comparison, the land needs for the "utilitarian" functions of working, shopping, circulating, hygiene, education, etc., are very small. Though this figure varies for countries such as England, the United States, Austria, Holland, and Israel, we would say that the similarities in the different countries are more striking than the differences. The amount of urban land needed per inhabitant is tending to become uniform throughout the world, and it is possible to assume that equality of recreational needs, wherever these needs are recognized, is the most important factor making for the uniformity.

No attempt has been made to measure the land requirement of modern townsmen for recreation on weekends or during monthly or yearly holidays. The larger the scale, the more intricate the calculation becomes. Such measurement would depend strongly, for example, on local climatic conditions, which might "compress" the yearly holiday period into a very few weeks of expected seasonable weather; on topographical and geographical conditions; on movability of urban population; on means of transportation; and on local custom.

The existence of great recreational pressure on land surrounding the metropolitan concentrations of population is well known, but no standards of land needs have been established. A hint comes from the Netherlands. In this densely populated and most intensively used land, natural areas amount to only 0.056 hectare per inhabitant (Buskens, 1951). That the Dutch complain of a definite lack of areas for week-end and holiday recreation within their country is an indication that the amount of recreational land has become insufficient for the needs of the population. In the United States the area of national parks, state parks, and national forests amounts to 0.6 hectare per inhabitant (American Society of Landscape Architects, 1954). And even that amount is, in the opinion of many American conservationists and landscape architects, wholly inadequate.

But are such figures of any real help in the calculation of regional needs? The safest assumption seems to be that the amount of land needed is very large. Surveys of demand vis-à-vis availability indicate that the need is still rising sharply. The present tendency seems to be toward a rapid increase in leisure hours and toward extending facilities and recreational areas accordingly. With the increase in population and the still growing congestion of cities, it seems that each new urban generation exhibits a stronger urge for recreation. On the other hand, motorization and construction of roads and airfields are making ever larger parts of the continents accessible far vacationers. To comply under these conditions with the theoretical needs for recreational facilities, huge districts – indeed, the whole of the regions surrounding large cities or even whole countries of high-population density – would have to be turned into recreational areas.

We may conclude that it has become impossible to provide sufficient land in the vicinity of most centres of population to serve exclusively for week-enders and holidayers. At the same time there is no way of suppressing the recreational movement into the countryside. Evidently, therefore, the quantitative aspect of the question of recreational land use on a regional scale cannot be seriously considered before going more deeply into its qualitative aspects: the motives for, and the means of, pressure of urban population on extra-urban land for recreation.

Recreational Pressures

The provision of recreational space in the house, the town, the region, and the country is essential for the harmonious conduct of urban life; it leads to a proper dimension of cells in which individual, family, and social life can take place, but it leads also to the securing of organic relations and harmonious transitions among these different levels of human association – the creation of a spatial rhythm of life. The daily, weekly, and yearly frameworks of recreation indeed exist in the strongest dependence on one another. Only if all of them can be provided for can the rhythm of individual and social life be satisfactorily maintained. The lack or inefficiency of one of them creates a direct pressure on the other. A slum is characterized not only by lack of space and obsoleteness of flats or houses but also by hordes of children and adults escaping their dwelling and filling streets, courtyards, and gardens whenever the weather permits. Since they do not meet in properly dimensioned squares or gardens but, instead, are compressed in narrow streets or yards, the nearness of one to another stimulates friction, quarrels, and hate among the fellow-suffers-proof of the fact that man, even urban man, needs a certain quantity of land under his feet.

A slum quarter, therefore, requires larger public gardens and squares, more public facilities of all kinds, than a healthy quarter; but, of course, every administrator and planner rightly prefers to invest money in the demolition of slum quarters and in their replacement with better houses rather than in the consolidation of slums by the establishment of facilities. We know today that town planning depends on and begins with the planning of the basic cells of community life – the dwellings.

In many cases, however, town planning also ends with provision for houses and minimal amenities within a street or neighborhood. The towns of our century have inherited an immeasurable volume of incompatibilities – social, aesthetic, technical, and educational. With a very few exceptions our larger towns suffer from huge deficiency in land areas for daily recreation, and none of the metropolitan centers meets the theoretical requirements for urban recreational land. We have to understand that this fact is the cause not only of poorly functioning towns but also of the heavy pressure of "land-hungry" urbanites on the rural countryside – for "Glasgow is a good place to get out of" (Abercrombie and Matthew, 1949, p. 130).

Similar to the process whereby erosion and floods result from the loss of absorptive capacity of the small particles of soil, the recreational movement to the country is the result of the obsolescence of urban dwellings and the lack of recreational land within the town. The recreation-searching masses turn into a "flood wave." We can assume that in many countries it is only the fact that large portions of the population cannot afford a holiday far from the city that have up to this time preserved large tracts of landscape.

On one hand, our civilization requires ever larger areas of recreational land, but, on the other hand, we are making the landscape ever more uniform and limiting its restful and beautiful parts by maximum exploitation of resources. The violent result is the invasion by townspeople into the rural surroundings of the city on fine weekends and holidays. Here a new conflict of interest between farmer and townsman has followed; the farmer looks upon the holidaymakers as pests – damaging crops, destroying fences, disturbing the cattle, burning the forests, and soiling the countryside. Indeed, a recreational area after the withdrawal of its visitors is a wretched sight. But the townsman, on his side, considers the farmer an egoistic tyrant who meets his visitor grudgingly and tries to prevent his short weekend enjoyment.

The better the economic condition of the average town dweller, the greater becomes the problem of recreational invasion of the countryside. Eventually, the growing numbers of holiday-makers begin to constitute a nuisance not only for the country folk but also for one another. Trying to return for a holiday to primitive conditions of life, people meet or "surprise" one another instead of finding solitude. Overcrowding prevails, just as within the city. Recreation here, like the trip from home to the countryside and back, is a nuisance, often more strenuous than the daily toil. Every big city knows those spots in its vicinity where recreation means only a change from an honestly artificial urban environment to a specially manufactured "natural environment" – a change from the difficulties of daily life to the difficulties of Sunday recreation.

For a large part of the population, recreation is spoiled when it does not offer them a chance to escape from one another. Even in the United States, with its comparatively large areas of wilderness, a conflict is evident between the desire to put recreational facilities and larger areas of land at the disposal of urbanites and the desire to preserve the natural countryside in its original state to make possible its solitary enjoyment by individuals and small groups (Feiss, 1950). The more artificial the urban environment, the larger the demand for compensation in indigenous landscape. But the most beautiful spots in a region are often kept a secret, because advertisement of them would mean their certain destruction by an influx of visitors.

The problem of recreational "inundation" of the countryside has to be tackled first of all inside the town by securing for the townsman the minimum measure of land he needs. A large part of his recreational needs thus would be met in his immediate environment, and the urge to leave the cities would be normalized. The whole character of outdoor recreation would be changed from one of flight from the city to one of harmonious movement of townspeople meeting their regional environment. But any such change for the better to be expected from town planning and development would not reduce the radius of travel for urban holiday-makers or restore the inaccessibility of rural and indigenous landscape. The same motives of social welfare which would encourage a community to enlarge its own recreational facilities would also induce it to prolong the yearly vacation of the average citizen and improve his chances of using that time for recreation outside the cities. In looking for a solution to the recreational problem, our main concern must be with regional development and regional design. We cannot expect a return to past conditions, and we are therefore compelled to turn our thoughts and energies to the comprehensively planned reconstruction of town and landscape as well as to the change ot attitude toward environment.

Methods of Approach to Recreational Planning

The beginnings of land-use planning for recreation lay with those romantic lovers of nature who demanded the preservation of indigenous or rural landscape in the name of God, the nation, or nature in general. Their approach was defensive, and their fight actually was for the salvation of this or that natural area and animal species from the impact of techniques and industry, and thus for its artificial separation from the landscape of modern civilization. For them the landscape was an indictment against our civilization, an offense against the wholeness of life.

We feel that theirs was a righteous cause; the rational arguments which they used to defend nature, however, were less convincing to businessmen and politicians. Investment in recreational facilities is by no means a good business proposition if such facilities are not intended for mass recreation. Nor could an expectation of greater man-hour production as the direct outcome of the influence of landscape on human health and vitality be substantiated. Arguments concerning the loss of income of local hotels, gas stations, and other small business were employed as a last attempt to preserve the integrity of the landscape (American Society of Landscape Architects, 1952), but expectations of short-term profits through exploitation of land for lumbering, mining, and power generation always proved much more attractive.

The truth might be that for conservationists the very existence of wild nature is the real issue. By advocating the part-time use of landscape as an amenity, they tried to influence a utilitarian society to co-operate in the realization of their lofty ideal.

Given the existence of such mercenary interets, it should be considered a most fortunate achievement that conservation societies and outstanding individuals have succeeded in many countries in preserving limited areas of wilderness as nature reserve or national parks. Even in these the fight for preservation against industrial or agricultural interests, on the one hand, and against invasion by holiday-makers, on the other hand, has to be vigilantly pursued. It is no wonder, therefore, that pessimism is widespread among nature preservation societies (Clarke, 1946-47). They understand that stretches of wilderness are becoming museum pieces – exhibits to show the coming generation what they have lost. The rate of deterioration of landscape is still much faster than that of preservation, and the prospects of accomplishing by preservation a finer environment are indeterminate.

But, while the fight of the conservationists is directed against certain basic symptoms of environmental change, it does not touch on the man-land relationship as a whole, on comprehensive environmental reconstruction. Positive goals of environmental health have to replace the defensive actions of conservators. As Patrick Geddes wrote in his *Cities in Evolution* (1949, p. 51), "The case for the conservation of nature must be stated more seriously ... not merely begged for on all grounds of amenity, of recreation, and repose, sound though these are, but insisted upon."

Out of the theoretical development of, and the still very limited practical experience in, regional and town planning, the most important conclusion to be drawn with respect to planning for recreation is the need for comprehensiveness. Land-use planning for recreation should be comprehensive in the geographical sense. For practicability, the interdependent recreational facilities of the house, the town, and the region have to be equally considered and provided for. The problem of recreational pressure on the countryside cannot be solved without providing first for the necessary recreational areas and facilities within the town. The same is true of planning for public open spaces in the town and the planning of individual houses and flats. On the hand, the most efficiently planned town, containing a full quota of recreational facilities, is still a beautiful prison if its regional surroundings do not offer the town dweller an attractive and accessible environment. Ample recreational facilities should confront man in all the different spatial frameworks through which he moves; the problem cannot be partly solved, because the very compression of recreational land use into an insufficient framework negates the possibility of recreation.

Planning for recreation in regions and towns should be comprehensive also in the functional sense. As far as possible, the environment planned for working, trading, circulating, and dwelling should be recreational as wel as utilitarian. To be effective, recreation has to be found casually in the factory at the hour of rest, on the way home, and at home. Vigilance with respect to the availability of recreational facilities should not be limited to a few zones or to the center of a city but should encompass the whole city – its houses, gardens, square, and streets, providing in one place nooks for individual seclusion and elsewhere for excitement and pleasure in a social context. Recreation would thus represent one of the elements composing habitability.

To the numerous extant formulations of the aim of planning we would, then add another: Planning aims at perpetuating recreation in all environmental frameworks. This implies that recreation should be part and parcel of the function of all land use and not only the destiny of specific chosen areas of land. It is the office of a planning program to turn town and country as a whole into a functional and aesthetically enjoyable environment.

When recreation is considered a part-time function of man, necessitating a specially treated, segregated environment, an awkward contradiction occurs: the more one plans explicitly for recreation, especially on the regional scale, the less satisfactory the result. There are several reasons for this difficulty. A planned natural or historic environment in holiday resorts cannot fulfil the longing of many vacationers to return to the lost rural or indigenous landscape. Neither nature nor history can be "designed." Attempts to do so have led only to the fabrication of ridiculous junk – ornamental "prettification" in a money-making atmosphere – but not to any true environmental quality. Also, such planning assumes on the part of contemporary men a sort of contentedness with the existence of "utilitarian" land areas, the inferior environment of everyday, for which, it is further assumed, part-time compensation can be had by recourse to a complementary artificial recreational environment. The dual existence of discrete ugly and beautified environment is thus perpetuated; it becomes the confirmation of the rupture between daily life and the good life, which is one of the marks of our big cities – the confirmation of a dualism which ought to be eliminated by planning.

Whereas the planning of separate zones for industry, through-traffic, and residence, as practiced today, seems to be in many cases a reasonable method, recreational zoning, as it is often proposed, may miss the very meaning of recreation: it is precisely the specialization of functions which upsets the equilibrium of man in the modern city and which should be balanced by variety – variety which recreation should provide. To become a true source of recreation, the whole of our regional surroundings has to be turned into an environment which provides for nourishment, occupation, interest, enjoyment, and health at the same time. Planning for recreation should be enlarged from compensatory or defensive zoning to planning for comprehensive purposes of higher environmental quality everywhere.

Summing up this short survey of the planning problems of the present recreational crisis, we present two statements:

a. It is impossible to provide for the theoretically needed amount of land for outdoor regional recreation if it is intended to be exclusively recreational land. Given the increase in world population, first call on land rests with food production, power generation, and industry – especially in the immediate surroundings of large population centers. Recreation, therefore, would have to be confined to the remaining "useless" wastelands, coastal and mountainous areas, or preserved stretches of indigenous landscape, wherever these happened to be located, and for as long as no economic importance was ascribed to them.

b. It is, however, not even desirable to develop a specific recreational environment on the regional scale for the part-time use of inhabitants of the large cities. Visiting such an environment may be a matter of social or erotic interest, of fashion or prestige, but it does not represent a true source of physical and psychological enrichment and renewal. The reason lies in the inevitable overcrowding, which, together with recreational specialization, should be considered as contrary to the essential recreational needs of metropolitan inhbiatants. From the point of view of quality of recreation, we have to search for areas of basically functional importance – areas of indigenous nature, agriculture, fishing, pasturing, lumbering, etc. – where recreation would represent one of multiple uses for such land.

Our conclusion, therefore, is that the crisis of recreational land use can be solved only by opening for recreational use the whole of a region. Nowhere should recreation be an exclusive function of an area; a landscape should be useful and beautiful at the same time – a resource of life and of its renewal.

But is it possible to expect the recreational need for rest and beauty to become the instigator of such a general reconstruction of landscape and environment?

Reconstruction of Landscape

There is an intrinsic conformity of aesthetic with functional qualities in an environment, and in this conformity lie all prospects for recreational improvement. To be precise: not all functions create environmental beauty, nor is all environmental beauty functional; but quality creates conformity between them. This was most probably sensed by those nature-lovers who maintained that disfigurement of landscape meant also the decline of our civilization and life. But, as long as mechanistic concepts of land as a food-producing substance prevailed, that feeling found no material "nutrient," and aesthetic and recreational values remained widely separate from reality.

Today the teaching of ecology, organic agriculture, soil science, and land-capability classifications are making conformity a scientific certainty. Now, indeed, "the case of nature conservation . . . can be insisted upon." The disfigurement of landscape is not merely a symptom but also one of the basic physical causes of cultural decline; it is the effect of a radical change in the relation of man to land and a new cause of human human deterioration as well. It is a source of vital aesthetic and recreational dissatisfaction and at the same time a source of deficiency in quantity and quality of food, water, climate, wood, and habitability of the earth. The recreational crisis is part and parcel of the general crisis of basic resources.

Though industrial developments are closely linked with the rise of the birth rate in many countries, the landscape as transformed by industry is incapable of providing the nourishment for an increased population over a long time. It is a landscape of man-made erosion and of declining fertility – and other ever mounting physical problems. All the emphasis is on maximum crops and high profits within the shortest time and for a price which is to be paid by future generations. The land can be interpreted as being functionally degenerate. To secure a permanent basis of civilization, a further step, one of environmental reconstruction, is needed.

In the shaping of tools, houses, and even cities we have learned the intrinsic relationship of material, function, and form, brought to high expression in handicrafts, architecture, and city design. Now, recent developments in biology have made us understand the natural processes to a degree where we begin to recognize our immediate power over, as well as our final dependence upon, the ecological functions. The outstanding importance of our new biological knowledge lies in the fact that it sets us at the beginning of new enterprises on a larger scale, which may be called "reconstruction of landscape," "regional design," or, as Geddes put it, "geotechnics."[1] This is a scientific enterprise as far as it it the observation and the emulation of nature's rule of return, and an artistic enterprise as far as nature leaves us the freedom, or even incites us, to express our developmental longings in the creation of higher qualities of environment.

The first realization of geotechnics – in the United States especially the Tennessee Valley Authority; in European countries the beginnings of afforestation and agricultural intensification, such as in Israel – as well as of the theory of landscape reconstruction, as developed in the last few years, indicates the changes in the cultural landscape to be expected: an increase in

[1] That Geddes used the term "geotechnics" is reported by Benton MacKaye in "Geography to Geotechnics," a series that appeared in *The Survey*, October - December, 1950, and April - June, 1951. New York: Survey Associates, Inc.

forests and wooded strips, an intensification and variegation of agricultural land use according to soil capabilities, terracing and strip cultivation, the following of lines of natural contours or soil qualities in the delimitation of parcels and fields, and the bringing to an end of the grid pattern of fields introduced by the land surveyor and the real estate merchant. There emerges a reallotment and redevelopment of whole rural countrysides, as begun in the Netherlands and in other European countries – a far-reaching reorganization of the treeless "food factories" or of the abandoned eroded fields into smaller fields bounded by wildlife strips.

The application of ecological principles of maintenance of soil fertility will lead in different countries to different landscape designs, because such application will be based on research into regional soil conditions and capabilities and human conditions. For many regions we can imagine as the result the creation of a pattern of freely curved wooded strips, traversing the plains in many directions, widening here and there into woods, running along streams and rivulets, and eventually connecting with the mountainous hinterland, where they would gain in width and finally merge into forests. The shady pathways, the rivers, and the forests of wildlife, for which people in many countries long, would again come to life – not because we should be ready to pay for recreation but because we should be obeying the scientifically recognized rules and preconditions for our permanent settlement and nourishment. Numerous planners have observed that in land-use planning on the regional scale recreation is always among the objectives "obtained . . . as collateral benefits" (Blanchard, 1950). Game preserves would be kept not because of the unceasing endeavors of conservation societies but because "the cover needed for watershed conservation [would be] . . . restored to the drainage channel and hillsides" (Leopold, quoted in Graham, 1944). A beautiful recreational landscape, as Sharp (1950), has pointed out, "arose out of activities that were undertaken primarily for other motives, rather than that it was deliberately created for itself."

We can imagine also an increase in planting along roads and trenches to avoid soil erosion and the planting of green belts around villages and cities to absorb the urban floodwaters, to minimize the range of influence of urban dust and smoke, and to create a harmonious transition of great recreaitonal value from town to country. Green strips may converge on the cities and even penetrate into them. Here certain new trends of town planning, which have already found expression in several countries, conform entirely with the large geotechnical principles of reconstruction. In earlier centuries the formally arranged private garden symbolized in a way the conquest and taming of nature by man. The free design of public gardens during recent decades has been the next step and may represent a memorial which the townsman erects in the heart of his city to remind him of the lost natural landscape. It is a condensed artificial landscape in which a large variety of plants, as well as rocks and water, often represents the natural landscape "in a nutshell." In many new towns, however, a new way of designing planted areas has appeared; these designs admit, without much artificial treatment, a wedgelike penetration of the surrounding landscape into the center of the city. In this way an extensive net of green pathways subdivides the town in a natural fashion into the residential neighborhood units; it represents the most attractive and convenient route of communication among places of work, homes, shopping centers, and friends, and it joins with sports fields, playgrounds, and schools. Here recreation has been truly integrated into the whole of the functions of urban life, and there is no longer a need for obtrusively specialized recreational facilities.

The new town no longer represents an isolated fortress, as in past centuries, or an agglomeration of houses alienated from its regional surroundings, as in the nineteenth century, but a regionally integrated nucleus of the landscape, from which open freely the channels which connect its center with the region and through which its lifeblood streams in and out. The function this pattern fosters and expresses may be interpreted as the mutuality of social and biotic life. The human communities of such a region can be strengthened only through the enhancement of its biotic communities. Its biological improvement, however, involves its aesthetic and recreational improvement.

Man has changed his landscape time and again. But all large-scale landscape design has been based on functional rather than aesthetic foundations. It may be expected that both "useful" and "useless" landscape will gain, by the new reconstruction of landscape, much of that "indigenous" character which is so valuable for recreation (MacKaye, 1928, pp. 138 and 169). But what does that indigenous character signify in this context? It would be superficial to explain it merely as a return to a primitive past. "Indigenous" should be interpreted, as MacKaye has, as a quality of past, present, and future. As appeared in a recent memorandum of the (British) Soil Association, "The primitive environment was better, not because it was primitive, but because the rule of the natural biological cycle prevailed" (Anonymous, 1955, p. 77). In the same way recreation would be better, not as an attempt to return to the past, but as a way to eternally desirable values. The indigenous character of landscape which may result from application of scientific methods would be a confirmation of the quality of our work of reconstruction. That landscape would be a realization of our aspirations toward health and wholeness.

Realization of Recreation

We began this essay by searching man for his needs and the landscape for its recreational resources; we found man's needs to be rising at the same time that the recreational landscape is deteriorating; only comprehensive regional reconstruction can restore the true sources of recreation. Now we have to look fo the human resources for this tremendous enterprise which may be described as *recreation of environment*. Our problem has become reversed, and it is no longer possible to separate "recreation by environment" from "recreational environment." Indeed, the very term „recreation" hints at this ambiguity: recreation means the revitalization of man's life by whatever circumstance, but it also means the restoration of life in man's biotic and physical environment. Recreating and being recreated – both are included in the original meaning of recreation, and, indeed, only in this double sense can it be realized.

We have dealt with the problem of recreation for the most part skeptically. As long as we are satisfied with *expecting* recreation from the environment, there is much room for skepticism. Hope begins when we deal with recreation in its active as well as its passive aspects. Such recreation loses the character of temporary compensation; it becomes a positive act of observing, enriching one's experience, widening one's interests, participating in the activities of communities, and developing receptivity for environmental qualities.

In our time we often meet the tendency to identify recreation with certain ways of behavior in free nature and in foreign places – a sort of planned emotionality and permanent enthusiasm. When we speak of "active" recreation, we aim not at the instigation of any such recreational enthusiasm but at positive purposes of recreation. Active recreation may become the voluntary preparation of the urban inhabitant for the geotechnical renewal of his region; it may be the first step – reconnaissance – in the long-overdue fight against soil erosion, declining fertility, and landscape devastation, aiming at the qualitative and quantitative enhancement of food-growing areas as much as at the habitability in town and country. This sort of recreation would serve the progress of regional survey of towns and country. As conceived by Geddes, it would renew our acquaintance with our regions, "rationalise our own experience," and prepare us for its planned change by widening our factual knowledge as well as educating us to a synoptic planning attitude; it would become "regional survey for regional service" (Boardman, 1944).

Wherever attempts at land reconstruction have been made, it has emerged clearly that this is a multipurpose enterprise, involving agriculture, water supply, power production, industry, transportation, and population movement and geared to residential as well as recreational purposes. To be successful, such an enterprise has to be undertaken by collaborating parties of different interests. The rural forces alone are in our time unable to accomplish the task. Urban scientific and technical achievements have to be fully applied to the country to bring about afforestation, dam-building, terracing, drainage, planting, reallotment of land, and construction. If repair of the man-land relationship were to become the essential content of recreation, the recreational return of the urban inhabitant to the land would mean the beginning of mutuality of urban and rural land-use interests and of co-operation in planned regional reconstruction.

We can now summarize by forecasting three stages of environmental development beyond those set forth at the outset – though these represent no certainties but only postulates:

4. Urban man should realize that, when he conquered the coutryside and created towns, he at the same time lost important environmental values. Forced thereby to search for his own recreation, he returns to the country. The more that industry and cities expand, the greater is the demand for recreation – but the greater also are the chances to realize recreation in its double sense by combined economic rehabilitation, social re-education, and physical reconstruction.

5. In the reconstruction of landscape, co-operation between town and country and among professions would re-create a fertile and habitable environment. It would be the greatest enterprise of planned environmental change since Neolithic times and the best act of social creation we can imagine. With the help of science, man reconstructs nature in its own image, which is at the same time his own best image.

6. Acting toward these puposes, man would rediscover the land as an inexhaustible resource of human recreation; making such discoveries, he would at the same time regain confidence in his own creative capabilities. Recreation would then become means and ends in one – and the earth, a better habitation.

REFERENCES

ABERCROMBIE, PATRICK, 1945, *Greater London Plan*. London: H. M. Stationary
Office. 221 pp. (In this work important beginnings have been made to cal-
culate the land needs of modern urban inhabitants.)

ABERCROMBIE, SIR PATRICK, and MATTHEW, ROBERT H., 1949, *The Clyde Valley
Regional Plan, 1946*. Edinburgh: H. M. Stationery Office. 395 pp. (Chapter iii,
"Open Space and Recreation," pp. 129-58, deals with the regional problems of
recreation of the Clyde Valley and helps to clarify the problem of recreational
land use generally.)

AMERICAN SOCIETY OF LANDSCAPE ARCHITECTS, 1952, "Selected 1951 ASLA
Committee Reports: Public Roads Controlled Access Highways, Parkways,"
Landscape Architecture, XLII, No. 2, 57-77. 1954, "Selected 1953 ASLA
Committee Reports: National and State Parks and Forests," *ibid.*, XLIV, 136-
37.

ANONYMOUS, 1955, "The Dental Health of Children" (memorandum to the British
Dental Association from the Soil Association), *Mother Earth*, VIII, No. 1,
75-80. London.

BENTHEM, R. J., 1952, "The Development of Rural Landscape in the Netherlands," *Journal of the Institute of Landscape Architects*, No. 25, pp. 2-9.
London. 1949, "Report of Documentation on Reconstruction of the Landscape
in the Netherlands." (Unpublished lecture.) 4 pp. (In Benthem's articles as well
as in his practical work the idea of the reconstruction of the cultural landscape
of a densely populated country comes to a clear expression; it means improvement as a multipurpose enterprise, comprehending useful and recreational
functions.)

BLANCHARD, R. W. 1950, "Master Land Use Plan for Crooked Creek Reservoir,"
Landscape Architecture, XXXVII, 140-41.

BOARDMAN, PHILIP, 1944, *Patrick Geddes, Maker of the Future*. Chapel Hill: University of North Carolina Press. 504 pp. (The life-story of the great biologist,
geographer, educator, and planner is at the same time a forecast of the science
of renewal of man-work and environment.)

BUSKENS, W. H. M., 1951, "Recreatie als Vraagstuk van Ruimtelijke Ordening,"
Natuur en Landschap, Vijfde Jaargang, pp. 21-29. Amsterdam.

CLARKE, GILMORE D. 1946-47, "A Challenge to the Landscape Architect," *Landscape Architecture*, XXXVII, 140-41.

FEISS, CARL, 1950, "National Park and Monument Planning in the United
States," *Town Planning Review*, XXI, No. 1, 40-56.

GEDDES, SIR PATRICK, 1949, *Cities in Evolution*. (1st edition. London: William &
Norgate, 1915.) 241 pp. (This book is the only publication of Geddes' numerous scripts on comprehensive planning. The more one penetrates into Geddes'
ideas and formulations, the wider appear the horizons of the future in thought
and action.)

GRAHAM, EDWARD H., 1944, *Natural Principles of Land Use*. New York: Oxford
University Press. 274 pp. (In this scientific work the importance of our new
biological knowledge for a comprehensive change of the landscape becomes
obvious.)

JACKS, G. V. and WHYTE, R. O., 1944, *The Rape of the Earth*. London: Faber & Faber. 313 pp.

LEOPOLD, ALDO, 1949, *A Sand County Almanac*. New York: Oxford University Press. 226 pp. 1953. *Round River*. Ed. Luna Leopold. New York: Oxford University Press. 173 pp. (Leopold's nature descriptions are of a lyric beauty. They reveal a deep longing for recreation of man in nature's eternal ecological cycles. His philosophical and ethical conclusions represent the most essential and concise appeal to man as a "biotic citizen.")

MACKAYE, BENTON, 1928, *The New Exploration: A Philosophy of Regional Planning*. New York: Harcourt, Brace & Co. 235 pp. 1950, "Dam Site vs. Norm Site," *Scientific Monthly*, LXXXI, No. 4, 241-47. (MacKaye's ideas of indigenous landscape, rural and city life, the flow of population and commodities, habitability, and active and passive recreation constitute most essential building stones of regional planning.)

MEARS, SIR FRANK C., 1948, *Forth and Tweed: Regional Plan for Central and South Eastern Scotland*. Edinburgh and London: Morrison & Gibb, Ltd. 180 pp. (The chapter on "Recreation and Amenity," Part V, pp. 141-49, hints at the true extent of the problem.)

MUMFORD, LEWIS, 1938. *The Culture of Cities*. London: Secker & Warburt. 530 pp. (The true importance of these interpretations of past and present environmental culture is revealed by the fact that they constitute a continous stimulation for thought, criticism, design, and action.)

NIXON, H. CLARENCE, 1945, *The Tennessee Valley, a Recreation Domain*. (Papers of the Institute of Research and Training in the Social Sciences, Vanderbilt University.) Nashville, Tenn.: Vanderbilt University. 22 pp.

RATNER, ROLAND, 1947, *Die Behausungsfrage*. Vienna: Gallusverlag. 120 pp. (The book reveals in a very simple and convincing manner the quantitative and qualitative aspects of urban land use.)

SHARP. THOMAS, 1950, "Planning Responsibility of the Landscape Architect in Britain," *Landscape Architecture*, XL, No. 2, 67-72.

TAUT, BRUNO, 1920, *Die Auflösung der Städte oder die Erde, eine gute Wohnung*. Hagen: Folkwang Verlag. 82 pp. (More than half of this book written by an architect consists of fanciful sketches of a reconstructed earth as a good habitation; the other half is an interesting collection of quotations from Kropotkin, Walt Whitman, Fuhrmann, Scheerbart, Oppenheimer, Tolstoi, and others – all on the subject of improvement of life and environment.)

UNITED STATES DEPARTMENT OF THE INTEBIOR. NATIONAL PARK SERVICE, 1943, *Recreational Resources of the Denison Dam and Reservoir Project, Texas and Oklahoma*. Washington, D.C.: Government Printing Office. 98 pp.

WRENCH, G. T., 1946, *Reconstruction by Way of the Soil*. London: Faber & Faber. 262 pp. (In this stimulating collection of Dr. Wrench's works, the agricultural, political, and economic problems of many countries are treated at different historic periods. The problem of quality of land use clearly emerges as the central problem of all civilizations.)

III

PLANNING WITH THE LAND

In the development of the man-modified landscape, it is a basic fact that the natural ecological communities of plants, animals and soil-life of a region do not support sedentary human communities, except in unusual conditions. In the ecological "constitution," man, like many other animals, is mostly a migrant. Since man first began to strive for a sedentary way of life, a large proportion of his efforts accordingly to be invested in orderly planned activities, aimed at changing his environment and maintaining fertility and habitability by continued interference – both destructive and constructive – in natural processes. To support sedentary communities, the ecological balance of natural regions first had to be disturbed and next to be artificially re-established on a new basis which favoured the existence of man. Villages and houses offering a protection against climatic and other exterior changes had to be kept in a state of habitability by continuous work.

To achieve all that, man had to discover the laws of growth, decay, and of the seasons, and to adapt his ways to them. Land was thus consciously "married" to human communities. In this "marriage" man's social contacts and economic operations resulted in a certain order of movement and rest on the land. This order "impressed" itself physically on the inhabited landscape and transformed it into a nourishing, protective, humanized region. In the man-modified environment biological, physical and artificially produced objects – woods, pastures, fields, water courses, villages – were set in certain mutual relationships: nuclei of settlement and processing appeared in central locations, while peripheral areas of production and supply were established and integrated by road and waterway systems. On such concentric regional structures the flow of energy, men and commodities moved back and forth, somewhat similar to the movements of a spider on its self-produced concentric web.

These, in very general outlines, are the underlying principles upon which the greater part of the cultural landscape has developed. The rhythm of movement and rest created a very important part of the morphological basis of regional structure. The concentric regional structure is manifested in infinite variations, following local and historical conditions, often leading to highly intricate city-centred patterns with primary, secondary, etc. nuclei of settlement. But for thousands of years the structural principle did not change in any essential respect.

The complexity of the concept "environment," or "region," as an area of influence surrounding a permanently settled place or places, seems thus to be implied in the very fact of settlement. It is a life-determing set of natural conditions and at the same time a man-modified space; it expresses, therefore, a relationship between human capabilities and bio-physical conditions. It is a shelter formed by the movements and energies of its inhabitants. It is "functional" in the sense that it fulfils the basic conditions of human life, but it also represents a definite framework in which human self-expression can develop and form the region as a communal artifice. Whereas for the nomad landscapes were transient facts, with temporary stations distributed along an extended route, the settled region is a permanently man-enlivened area. Whereas repeated migrations from station to station involve great losses of energy, settled man tries to achieve an intensified communal life by the planned use of his materials and energies, by what Emerson called the "true economy of saving upon the lower planes of life, to spend on higher ones."

If we consider this traditional rural landscape to be a superior environment, the reason it not that it is more "natural" than our own surroundings, but that it fulfills all of the above-mentioned functions and at the same time is ecologically balanced. The concentric pattern of human settlement and land use may also serve us as a sound archetype when we try to restore the disturbed man-land relationship by Regional Planning. But once we come to recognize it as a result of a certain rhythm of rest and movement, we must ask ourselves if that rhythm has not undergone essential changes; and if human life values in our time are not intrinsically connected with a new order of movement and rest.

We can discern in history a trend towards a more complex settlement structure and a greater mobility, extending the range of human activities and conferring a greater degree of independence from the land. It is manifested at first in the appearance of migrant artisans and traders, then in the foundation of non-agricultural settlements and occupations, next by occasional plunder of commerce, finally by a highly developed technology, producing self-propelled vehicles and increasing the interchangeability of resources, conditions and experience. In a way only coplex setlement structure containing towns can be viewed as the result of the balancing of sedentary and "nomadic" trends. It is generally true that the latter has often led to misues of resources, exploitation of populations and widespread misery; whereas man the settler, as several authors have observed, succeeded in procuring a peaceful co-existence with the modified ecological communities for many thousands of years. Yet our work as planners cannot succeed unless it suits a wide variety of human ways of life. Any dogma about the way man should be related to his environment must be revised by knowledge about man. The attitude upon which our plans are based should be broad enough to allow for both the trends of sedentary settlement and mobility. Human development can be interpreted as a process in which man's contacts with and participation in the life of the surrounding world is ever enlarged and in-tensified. We should therefore take account of the fact that man's desire for such intensified participation extends both in depth and in width; it may induce him to adapt the "vegetative" trend to belong somewhere and take root there as a settler, or the "animal" trend to move about as a migrant agent of change. As a matter of fact, an ever growing segment of the world's population is adopting willy-nilly the latter kind of life. The high degree of mobility which has lately become possible for the mass of the population in many countries may be more than a passing phenomenon. As planners we should take into account the possibility of a more dynamic rhythm of movement and rest extending over many parts of the world. We should how-ever ask ourselves what are the aims of the new molibity, and in what manner can they be expected to improve and enrich the life experience of men?

In planning the future human environment, one needs an open mind to the possibilities of human development, and open eyes for observing the human environment. Recently some attempts have been made to "idealize" the environment created by contemporary civilization. The emerging land-scape of mechanized land exploitation, highways and industry has been des-cribed as the best one possible perhaps merely because of the fact that it is new. The reason for this readiness to idealize such a landscape may lie in the absence of any accepted scale of environment values, which would help us to interpret the changes we observe in town and landscape. We may sense vaguely what a good habitable region can be, but we have lost the under-standing of its inherent values for man.

The traditional rural region, with its balanced urban centers, was a com-prehensively used environment. It served its inhabitants as a place of work, play, rest, social contacts, exchange and celebrations. In this sense it was fully "humanized." At the same time it was distinguished by a permanently conserved fertility and water cycle, and a health-giving landscape. In most countries these human and ecological landscape values have been largely destroyed or are being currently destroyed. Regional landscape is turning from a humanized environment into an object serving specific one-sided purposes. Contemporary changes in the landscape are caused by a city-centered and technologically highly equipped mass-movement. For the first time since men abandoned the historic forms of nomadic life, the pattern of our regions is being decisively conditioned again by the forces of transit.

There is, of course, a great difference between regional deterioration in areas of industrial over-development, and the over-populated, exploited and starving areas of raw-materials production. But the common feature of both is that it is not a single specific technological, commercial or political development which causes these changes and which requires adjustment; it is rather a rapid succession of often conflicting new factors, methods, and movements, which disrupts all possibility of attempts to establish a man-land relationship on a new level. Energy derived from coal, petrol, electricity or radio-activity; railways, cars, planes, and helicopters; tractors, bulldozers and harvesters; colonial, industrial, commercial, residential or recreational expansion – all of these factors condition the peripheries as well as the centers of regions in different and un-coordinated ways.

As a result landscape and urban development are marked by basic incompatibilities, such as the concentration of business as against the decentralization of residential areas, the individual car's traffic requirements as against the maintenance of orderly city-life, the increasing demands for recreational areas as against the deterioration of the recreational landscape, etc. Aesthetic dissatisfaction is greatest where a wholesome environment is most needed. A new kind of a formless mass movement of men, commodities and energy is expanding over the regions, and landscape has begun again to be a transient fact. Under this impact, the cultural environment is rapidly approaching a point of degradation which renders the landscape desertlike in appearance and effects. It may assume the form of the "steppe of culture," as the Dutch call the open landscape of treeless rectangular fields which began to emerge at the end of the last century; or the recreational areas and resorts where the very recreational values are destroyed by mass use. Landscape as a habitable and recreational environment is being rapidly overrun and destroyed. But deterioration is becoming apparent also in the social and architectural desert of suburbia, spreading over densely developed regions, or in the new standard international landscape of highways and airfields.

In highly developed industrial regions this transformation of landscape is accompanied by a change in the ideas about landscape. A formulation of the values and contents currently attributed to the cultural landscape reveals in general a rather ambiguous attitude: landscape is considered on the one hand as a zone of raw materials production, a huge factory for wood, food and minerals, to be populated as sparsely as possible by a working population using machines and fast vehicles. Landscape is viewed also as an empty space lying between points of interest, the cities – or even as a spatial obstacle to close socio-economic and cultural contact between population centers. As such, it is to be bridged over and "nullified" by ever faster communications. But at the same time the landscape is also imagined and searched for as the "natural" idyllic environment, an object for nostalgic dreams about older and better days.

With the prevalence of such ideas, environment as a life-favoring fact may be quickly approaching its end. The disastrous consequences can be delayed, thanks to the inherent adaptability of man, but in the long run these environmental deficiencies are bound to become unbearable. The substitute environment provided by the cinema, illustrated journals and television, or by the fake-environment of many holiday resorts, highlight the attempts to escape from the reality of our own time and space. We have lost control over the development of our own regional environment which takes place through the entirely uncoordinated application of new methods for achievements of numerous unrelated aims.

The land-planners' first task may be the re-education of their eyes and minds to perceive the landscape by using all the contemporary means of observation. This is fundamentally the function of the Regional Survey. To become familiar with the extremely complex situation and underlying conflicts of a changing region, we must first of all use scientific methods of analysis. This will provide us with necessary information about the demographic structure of a population, its sociological and economic relations and problems etc.; other scientific disciplines will enable us to map and describe the region's land, water and other resources, climate, etc. The scientific approach is a *sine qua non,* as the ingredients which comprise the regional situation cannot be revealed in another way, and because the distance between the planner and the region, as well as the scope of the work will always necessitate application of indirect methods of observation. This part of the survey technique is well developed today. In fact, so much is being done in such specialized studies that we are in danger of forgetting the true meaning of the notion "Survey," which implies a view from above presenting an integrated picture. In addition to the indirect approach, we then need a great deal of direct experience and observation of the regional life processes, and to complete the picture, the view from the air. We are perhaps only in the very beginning of being able to interpret properly what we observe from the air, and we might discover in this way important relationships between human, biological and physical facts and processes in a region. The most significant part of the regional survey will be the combination of the results of observations from these different angles. It becomes obvious that the Survey has the double function of providing us with information and of educating us in comprehensive thought and view. Both of these make a vital contribution to the mental preparation for planning.

The balancing and elaboration of these survey methods is a task which must be tackled anew in each region. A region is so much a unique set of facts and conditions that it would be a contradiction in terms to work out a detailed method of Survey to be applied in several places. A wasteland to be colonized calls for different methods than an over-populated, destitute rural region or a metropolitan conglomeration. Even such categories are much too general for the application of a unified methodology. A great amount of previous information and imagination are needed to elaborate a suitable method of survey which will result in more than a pile of unrelated facts. The more we succeed in creating a plastic image of the particular regional relationships in space and time – of the regional character – the more we can hope to achieve by planning humanization of the environment.

The most important fact about Regional Planning is not that it represents an intellectual advance which helps us solve our problems in a better way than was previously possible, but that in our time any prospect of achieving orderly regional relationships and functions depends on the application of rational and comprehensive Planning. In the past the structure of rural-urban regions resulted from the obvious mutual dependence and the subsequent organic integration of population in easily surveyable socio-geographic units. In general, this was accomplished by trial and error. Functionally and aesthetically, these structures are perhaps higher achievements than all the possible results of our Planning. It is the scale and complexity of contemporary environmental problems which preclude, for moral and political reasons, any but a planned regional development. The "organic" solutions must be accomplished by us through comprehensive organization. Regional Planning, therefore, represents one of the characteristics of the new phase in environmental development.

Planning will be based on an analysis of the regional emergency which calls for it. An essential part of the emergency can be represented graphically in charts and reports describing socio-economic relationships, land-use problems, transportation, communications and energy flow. The charts will indicate the specific regional rhythm of movement and rest, and the supra-regional, perhaps universal, influences bearing on the life of the region. The proposals for new environmental structures will attempt to integrate regional and supra-regional trends in a unique new order. Regional Planning does not mean planning for secluded regional units, but for regional and inter-regional relationships to be balanced in carefully defined area units. The regionalist attitude might spring from a universal humane concern, yet the only way to concretize such a concern lies in its application to specific places, communities and situations, and in the fostering of a specific environmental character.

Generally speaking, the new regional rhythm of movement and rest will be oriented towards both expansion and intensified coherence. As population and requirements increase, areas of settlement, production and recreation will have to be enlarged and the communications' network replanned in order to facilitate contact between communities and places. At the same time we also plan for interregional contacts, and of special interest are those between highly developed and poor regions. The problem of the new regional environment assumes one form in areas of affluence, and quite a different form in the poor overpopulated regions. Typical problems of affluent areas are metropolitan conglomeration and technological development – especially automation in industry – leading to a greater leisure for the masses. This in turn, requires the creation of a more accessible and hospitable regional landscape. The other type of region is generally characterized by the lack of basic resources, housing and machinery. For such regions technological advance and skilled manpower are not originators of new problems, but urgently needed means to secure physical survival.

These types of regional problems exist concurrently but are dealt with separately, as if there were no connection between them. The inequality between them creates a tension which endangers the existence of the greater part of humanity. An analysis of these interregional problems must lead to important consequences in the Planning of each particular region. It should have a considerable influence on the aims of population movement, the settlement structure and regional reconstruction in each place.

By regional and inter-regional analysis we shall come to realize how much we still depend everywhere on the environmental order, though the character of this dependence is not the same as in the past. The future does not lie in a mere return to the focal regional patterns. New facts are modifying the substance of regionalism. The new regional structure must provide a good home for communities and individuals, it must suit the long-range, efficient and intensive use of the land as a source of food and raw materials, but it must also suit the trends and possibilities of mobility, interchangeability and increased interregional contacts. The last point, in particular, must modify our ideas of regional structure. Whereas the city has represented, until this day, a separate focal complex which "sponges" one-sidedly on the land by an intricate network of roads, rails, pipes and wires, the population's new mobility requires a much evener distribution of the values of habitability and hospitability over the regional area. We have to plan for the interpenetration of town and country as areas of residence and movement, but even where the country is not permanently inhabited, it will become a frequently used habitat. To accomodate the new rhythm of movement and rest, the whole of a region would represent a space of work, habitation, transportation, social contact and recreation – in other words, a comprehensively used environment. The new type of mobility should involve not the creation of a transient environment but the modification of the rural areas to receive the movement of the return to the landscape. The existing ribbon developments of technical, commercial and amusement facilities along the great motorroads must be transformed by Regional Planning into a landscape development in depth. The landscape must not be excluded from the highways; it should positively accompany the travellers and be converted into an accessible, partly habitable and hospitable environment for the whole of the regional population.

Land Planning should become an attempt at balancing a measure of environmental belongingness with a measure of free mobility, at shaping a rhythm in the transition from movement to rest and vice versa, and at establishing by the introduction of environmental changes a meaningful and valuable relationship between men and their landscape. That relationship must find an expression in landscape quality. But landscape quality does not result automatically from regional land use schemes; it requires the development of Regional Planning into Regional Design. It is not the sole function of environmental planning to define the areas for residential, industrial and other uses. Rest and movement of men in the new landscape should not only be "channelled" by utilitarian schemes which define zones of use; they must be integrated in the landscape by proper design. Not the city, but the regional landscape must become modern man's much frequented "home." Consequently we need an architecture on the regional scale.

So far land-use planning and landscape architecture have taken place in complete separation. Landscape architecture had only a tenuous relation to the contemporary function of land, while land-use plans took no account of the need to form a landscape character. But we cannot continue to regard landscape beauty as connected only with indigenous or traditional rural – and today mostly destitute – landscape. On the other hand there is no way of beautifying a regional landscape except through the planning of its social and biological functions. Just as in architecture the functional syntheses found for a variety of programmatic requirements have to be elevated to functional quality; so the useful and the beautiful must be fused.

We are still a long way from the realization of Regional Planning on a scale adequate to the environmental emergencies in many parts of the world. We are even further off from a realization of Regional Design on a functional basis. Landscape changes have to be planned, propagated and actively promoted by an urban population which recognizes that its expectation from the environment as a nourishing, health-giving and recreative factor can be fulfilled only by the active recreation of the regional landscape. Planning must modify its ecological balance, and its habitability and hospitability should be established on a new level. The problem is how to turn the merely passive dependence of the urban population on the landscape into an active interest in fostering a cultural environment. The difficulty is that in our present stage of civilization, such interests as long-term ecological stabilization, environmental health and the kind of recreation which is not merely a form of escapism but a life renewing activity, are not sufficiently attractive targets in themselves. We are very modest in our demands from the environment. But man's adaptability to environmental conditions is only apparently unbounded. Environmental change on a large scale will become a convincing ideal when the long-repressed biotic and psychic urges of human beings for direct vital experience and contact in human relations, nutrition, and environment will prove too strong to be stilled by the various substitutes for reality which technology can provide. That turning point may re-orient our functional and aesthetic interests. The environmental change will come only if actively pursued, not as a result of "historical necessity."

Planned environmental development does not mean renunciation of technological achievements. It would be disastrous if our dissappointments with the use of technology would lead us to its negation. Under the conditions of population increase and enhanced inter-regional dependence, technology is the most vital means of achieving any improvement in human living conditions. Having moved far away from traditional settled conditions to semi-mobile ways of life, the integration of technological and environmental development has become immensely more difficult than in the past, but at the same time, it appears, the effort has become more worthwhile.

It is the function of Regional Land Planning first to open our eyes to environmental values, next to prepare for the immense work of re-creating the human environment, which, if we regard the future with hope, will be one of the central tasks of the coming era.

IV

THE PLANNER GEDDES

It is a century since Geddes was born, and twenty-two years since his death.
Yet the discussion of the theory and practice of Town and Regional Planning, which he initiated, is as alive as ever; instead of fading out, it would
appear that the debate has been put on a much more advanced plane by the
experience gained during the last twenty years in the building of new towns
and the reconstruction of towns and regions in many parts of the world.
What role do Geddes' ideas play in these recent enterprises? Their planners
and builders do not deny Geddes' historical importance as the foremost ini-
tiator of the new Planning Movement. But opinions differ as to the extent
his ideas have withstood the test of time. This question deserves further discussion. As the recent developmental enterprises in England, The Netherlands, U.S., Israel and other countries constitute merely partial achievements,
they clearly harbor the dangers inherent in any partial realization – the loss
of the grand directing conception. Recent experience has shown the danger
that, by paying too much attention to the details of Planning, one may lose
sight of the broad objectives of Planning as such. At the present stage, therefore, we may find particular value in a redefinition of Geddes' idea of Planning; it might help us to examine and re-consider the course Planning is
taking nowadays.

We cannot cite for this purpose any "Geddes Doctrine" providing final
definitions of the elements of Planning. Geddes did not develop the theory
of Planning in a single systematic study. He left articles, lectures, pamphlets
and innumerable reports on the subject of Planning, out of which only a
handful have been published to date. Most of them were written for specific
practical purposes. They constitute, in effect, illustrations of an unwritten
doctrine, accompanied by brilliant comments, ". . . notes written as it were
on the margin of his thinking," as Lewis Mumford puts it. Neither did
Geddes leave a legacy of new towns or regions built or constructed according to his plans, and which testify to his ability and to the soundness of his

doctrines. He worked at planning in many places, but his plans have been only partly realized. It is very difficult today to identify in Edinburgh, India, Cyprus, Tel Aviv, Montpellier, etc. what has been done by Geddes and what has not. North Tel Aviv, for instance, has retained in spite of many subsequent changes the basic characteristics of Geddes' street planning, especially in the quiet side streets such as Arnon, Rupin, Zola, etc. However, contrary to Geddes' proposals, these areas have been built up too densely and have little in common with his vision of a garden city, consisting mainly of small houses.

Any attempt to evaluate Geddes' achievement encounters the same difficulty: he did not leave after him "monuments," either of written doctrine or of buildings perpetuating his thought and action. To understand Geddes one must reconstruct the essence of his ideas and outlook on life out of his scattered writings and works, memoirs of his pupils and acquaintances and biographical works about him. These inquiries bring to light the fact that it is Geddes' whole life which gives evidence for his planning doctrine. In a most surprising manner, we realize the logic of the idea of planning by a study of Geddes' life. The sciences and theories with which he dealt consecutively as well as the whole of his social activity and his characteristic approach to practical and theoretical problems are identical in principle with the sections constituting the wide field of Planning. At first glance, Geddes appears as of an inconsistent turn of mind, as a man driven by excess of vitality into a medley of subjects and affairs. In effect, though, each stage of his life is a part of the proces of elucidation and realization of the idea of planning, which emerges at the end of his life, i.e. in our time, in its full scope and vitality.

Geddes' course led from direct observation of natural phenomena (highly varied around his Scotch birthplace), to the study of biology. His early discoveries in the twilight zone between plant and animal life, and his major share in the development of cell theory, marked him for a time as a definite successor to Darwin and to his teacher, Thomas Huxley. Geddes was chiefly interested in the relationship of the living organism to its environment. The life of a plant or an animal is not delimited by its body, but also covers a certain neighborhood, building it and being built by it. It is continuously in mutual relationships with its environment. Geddes' thought approximated closely at this point of the ideas of his friend, Peter Kropotkin, even before he met him. From the start Geddes included human life within the scope of this concept; the relationship of a man to his environment is one chapter only within the wider subject of the relationship of the living organism to its environment. In this manner, Geddes was led by the study of biology and later, geology – to an interest in the social sciences, to anthropology and economics on the one hand, and geography on the other hand. Dwelling on the state of the relationship of mankind to its environment, Geddes recognized the existence of a grave crisis. This crisis was exemplified by the environment city dwellers in Scotland and England lived in. It was, in fact, the environment of all cities and countries influenced by the 19th century industrial development, which created slums, rural degeneration, dusty, smoky and crowded industrial areas devoid of light and air and harmful to men and to society in body and in spirit. Such an environment hinders the full development of man's potentialities and threatens him with degeneration.

When we examine Geddes' fields of research and his public and organizational activity after being appointed Professor of Botany, we find him occupied in highly varied interests such as biology, sociology, economics, geography, statistics and pedagogy (especially the problem of University education.) Geddes analyzed them as mutually complementary sciences. He was seeking for an avenue leading beyond the highly specialized sciences, capable of unifying them into a single, comprehensive and useful science of life. He saw a necessity for a significant widening of a student's knowledge to eliminate narrow specialization. Geddes himself mastered many sciences, especially those referring to man's behavior in his environment. An Economist unfamiliar with the elements of biology, in Geddes' opinion, would not know anything about the elements of economics, as he is unacquainted with the elements of human life. A geographer ignorant of economics, anthropology, agriculture and sociology has only a knowledge of external dry facts but not of essences. This wider scientific horizon – being a *sine qua non* for each planning mind – makes possible the intermeshing of sciences in a certain consistent order and the creation of connections among them. It is a necessary condition for the transformation of the sciences into a tool serving the good of humanity.

Geddes had a peculiar talent of relating his comprehensive intellectual syntheses to a specific reality, and to illustrate them by concrete cases in concrete locations. He could also give a meaning to the general visual view of a town, a river valley or region by giving it stimulating, surprising, and convincing interpretations. His synthetic theoretic insight complemented his empirical impressions and vice versa. This synoptic method of his had a strong influence on his pupils and successors: on one hand this method was an expression of a most vital personality and on the other it brought his ideas ever closer to the most important problems of our time. This approach, as well as Geddes' scientific work and his search for a synthesis connecting the sciences, achieved its highest expression in his definition of the factors of folk, place and work as basic in determining by their mutual relationships the essence and form of human culture in every place and in every period. This definition was a development of the French sociologist Frédéric Le Play's similar definition. Geddes, however, considered this trinity of concepts only as the further development of the basic biological factors – organism, environment and function. Mutual relationships develop between the folk, place and work factors; they influence each other quantitatively and qualitatively, creating thereby a characteristic economy, influenced by a specific community and geographical location – or creating a town determined not only by climate and topography, but by economic relations and the human condition as well, etc. With Geddes' interpretations, all anthropological, economic and geographical science becomes related to the specific circumstances of Space, Time, and Man.

With clarification of these ideas, Geddes returned to concrete realities. He used them to lay a basis for the Survey. His Survey method, covering towns, regions and countries, relied both on science and on direct impressions. It constitutes a highly interesting development of the scientific approach, as it brings about a collaboration of several scientific branches for the analysis of a common concrete problem. The Survey is a training of synthetic thought and at the same time also a training of the eye to look directly for the essence of phenomena. This wide basis of the Survey has been largely forgotten nowadays, although the approach to the Survey also determines the approach to planning. If the Survey is of a merely technological or bureaucratic nature, as it often is, neither can planning overstep these limitations. A survey of a town or a region does not mean merely a listing of facts, but an insight into the essence of social interrelations, as well as the relations between the society, its sources of livelihood and its habitation. As the Survey progresses, the planner begins sensing clearly the characteristics of regional life. There is nevertheless no automatic transition from Survey to Planning. Planning does not grow gradually with the progress of the Survey, as the Survey, even when most perfect, merely establishes existing facts without creating the plans needed to effect the transformation of the facts. The main task of the surveyor, therefore, is not to widen ever more the scope of his survey, but to make a wise selection of the factors whose investigation should yield the richest harvest.

A crystallization of the plan of the operations calls for a will to achieve positive aims and a capacity for architectural creativity. Geddes mentioned this fact in a simple manner in a conversation reported by the American planner, Benton MacKaye: "Geography (and we can assume the same for the Survey too) is a descriptive science; it tells what is. Geotechnics (Geddes' own term for Planning – the architecture of larger space?) is applied science; it shows what ought to be." To know how to make the best use of the earth, ideals and a description of the desired goals are essential. In this manner Geddes made a basic distinction between science and social practice, between the Survey and the Planning, thus making a major step beyond contemporary determinism. The betterment of human life involves not only a knowledge and an interpretation of what is, but also an expression of the will of society as to what ought to be. The experience gained in the building of new towns in Britain, the reconstruction of Dutch agriculture, and the agricultural settlement of Israel seem to prove the importance of the social ideal as a necessary element in any successful settlement project.

Planning, for Geddes, represented the unity of practical knowledge and skill with social ideals for a positive use of natural conditions and human potentialities. Planning, nevertheless, is only a part of a threefold social action based on the recognition of the factors of society (or folk), economics (or work), and town and region (or place). The action consists of education to good citizenship, economic reconstruction adapted to the peculiarities of folk and place, and the planning and developmental activity aimed at changing the human environment. These three activities should be coordinated with no clear dividing lines between them. The transformation of the environment is a major architectural creation, therefore a self-expression of the society performing it, and the result of the way the society is educated. The execution of such an enterprise, is, however, at all times an integral part of a plan for economic rehabilitation. Society, by the mere fact of participating in the developmental enterprise of a city or a region, does educate and transform itself, while the conditions of the new environment created by the social group exercise in turn an influence to transform is. Geddes' conception admits no separation between planning, economic reconstruction and social education. A glance at one of the most succesful developmental projects of our time, the Dutch North East Polder, would prove that the secret of its success is just this unified execution of the three activities.

This shows to what an extent Geddes' conception of planning transcends the usual contemporary technological practice. Comprehensive Regional Planning becomes here a social expression embodying both practical and lofty ethical objectives. Let us look into the eventual conclusions to be derived from the consideration of the three factors – folk, work and place – in the developmental planning of a country like India. Such an action would be based on the peculiar culture of the Indian people. Even when such an operation is conducted by European planners, it would not constitute a merely technological, industrial, commercial or administrative operation – a transplantation of western civilization to the orient – but the development of the very potentialities and beauties of the country itself. This is what Geddes tried in India. He based his planning policy on a comprehensive theoretical preparation, with due respect for regional culture and the practicability of the development project.

Geddes' life is no less instructive in this respect than his theoretical writings. Geddes tried out his theory in practice. In 1897 he went to Cyprus to assist in the planning and execution of the settlement of Armenian refugees who flowed to the island from Turkey. His proposals and activities can serve as an example for the solution of similar problems that came into being recently in many parts of Asia. Geddes demanded that improvement of conditions should come as a result of long term investments, and not by philanthropy or heightened commercial activity. Aid should be "geotechnical." The solution of the refugee problem, as he envisaged it, lay in the intensification of land cultivation. Geddes' activities, therefore, consisted of planning, organization of financing, and settlement of the refugees in the agricultural farms and villages which they were directed to build with their own hands, the establishment of an agricultural school, the discovery of water sources (here he found use for his geological knowledge), the development of plantations, repair of irrigation networks, etc.

In the light of his experience, which was apparently highly successful, Geddes interpreted the history and problems of Cyprus and the whole Mediterranean area as the problem of the relation between man and water. "Solve the agricultural question and you solve the Eastern Question ... but politicians, diplomats and administrators are all agriculturally inept. The futility of the efforts heretofore is but the common urban incapacity to govern agricultural populations, to deal with rustic questions." Few could equal Geddes as far as drawing conclusions from knowledge and personal experience goes. The decade he spent in India in planning works in all parts of the country, as well as in educational and civic activity, serves as a proof. Our advance towards lending constructive assistance to "backward countries," towards practical geotechnical UN activities and technical aid, can be interpreted as a direct continuation of the path Geddes began to follow sixty years ago. "Peace is not fundamentally a question of high concert, conference or arbitration ... it is a question of industry ... mainly of agriculture ... *il faut cultiver son jardin*. That is the hygiene of Peace."

Regional reconstruction was viewed by Geddes as both a regional and an universal problem, and it seems that recent experience proves the correctness of his convictions, and makes the understanding of his achievement easier. An ever-growing understanding of Planning was considered by Geddes as one of the vital contemporary steps toward overcoming the limitations of our "pecuniary culture." This hope has not been realized as yet. Our era can even be considered as a period of reaction against planning. In many countries planning has been perverted into a merely technological operation, subordinated to the demands of industrial development, to general shortsighted projects in communications, engineering, etc. In many places, Geddes' comprehensive conception has been all but forgotten.

Let us hope this is merely a temporary relapse. Today it seems obvious that giving up the idea of reconstruction by means of comprehensive rational planning signifies despair as to the future of human society. One of the necessary steps to finding solutions to the critical contemporary problems is our rising to understand Patrick Geddes' idea of comprehensive planning.

V

SOCIAL TRENDS IN REGIONAL PLANNING

For nearly three years there has been in existence in Israel a very intensive Planning activity on the national, regional and local level simultaneously. Many development problems, urgent but different require efficient solutions, which have to be found in the shortest time. Normally it would take months or years to prepare the large schemes and proposals with which we are occupied. But owing to the great pressure there is generally no other way for us than to elaborate these schemes in the course of a few weeks only. So we do our best to use our knowledge of the country and its population and the experience gained in the field of Planning in other countries; we consult many institutions, offices, organizations and individual persons and often we have to rely on quick decisions and even improvisation. In the mean time the new immigrants arrive in tremendous numbers; they settle in all parts of the country; farms, villages, cities, industries and many public works are being constructed. A vast process of settlement is in process of realization, facts are being established, but only very seldom do we find time to consider our fundamental Planning attitude, the broader consequences emerging out of our quickly prepared schemes, the intrinsic ends we are aiming at, the human and social tendencies, which form the general background of our work.

The real scope of this notion "Planning," is much broader than indicated within the confinens of our daily Planning work; but the limits put to our actual influence should never be lost to sight. It would be wrong to assume that it was our task as Planners to provide a design for our future in this country, or "to define the cultural aspect of the Israeli man . . .," as expressed in a recent article here. Our task in this field is very modest – and this perception is one of the very clear results that emerged during the short period of our work. It is outside our objective to give substance to the life of man and society. It is for us only to contribute to the creation of a frame of life, which fits the human and social needs of existence, which paves a way towards the development of an "art of living." But if we shall succeed in gaining really a new "art of living" – that is a question, which will be decided on quite another level than that of the Planners. Even so Planning in our time has a deep historic background, roots as well as tendencies. Planning is not a social movement but an expression of a social trend. I believe, that this seminar provides a fitting opportunity to discuss the roots and tendencies of Planning and to criticize some actual proposals and developments in the frame of our Planning work from this point of view.

Planning is a social activity. Neither the scheme for a small quarter nor yet the plan for a normal One Family House is merely the result of the brilliant brains or gifted hand of one individual Town Planner or Architect. The scheme for a town or a region means even more the coordination of social needs and tendencies; and their "translation" into an efficient Planning scheme demands manifold activities—contact and negotiations with many institutions, and the team work of a group of Planners. Now this is more than a mere administrative or technical job: there would be no chance to achieve that coordination or collaboration, if it could not be based on the perception of some common need. We have to consider that Planning is not that kind of social activity which exists at all times and under all conditions. We learn from experience that Planning is a reaction to a state of emergency in a community, an existing or menacing emergency; thus Planning may become the expression of a positive purpose of life or a social affirmation of life, striving to overcome threatening dangers. And the analysis of the character of that emergency, which asks for Planning, will lead to the synthesis: the trends of Planning. This analysis forms the basis of the Planning policy as well as the basis for the collaboration of the Planners.

At the time of the last elections to the Knesseth, Mr. Sharon proposed in jest the establishment of "The party of Planning." It seems to me that this has some serious basis. It is my impression from the contact I had with active Planners of several countries, that this "party" is even an international party: the criticism of our civilized society and perception of its immediate dangers for the human condition are common to those, who in our days have found the way to Planning. You may even say that the Planners of many countries understand each other easily with the help of their common "language" – the Planning language.

What is the character of such dangers, which induce a community to plan? The fact that the Planning of the medieval and renaissance cities of Europe and of this country was based on needs of military defence may serve as a very elementary example, of greater importance are the planning projects and Laws of many countries in the Near and Far East. Here the installation and maintenance of the irrigation system became a question of life or death, which influenced deeply all public activities, so that one author even called the Oriental society: "The Hydraulic Society."[1] Even in our day the operation and maintenance of the irrigation system exerts the greatest influence on the physical and social organization of, for instance, the "Mormon Republic" in the state of Utah in the U.S.A. The need for permanent common action against inundations and sea floods forced the communities of large parts of the Netherlands a thousand years ago to plan their regions. The whole layout of the road network, the location of villages and towns, planting of trees etc. – but also the administration and organization of the communities and the regions – had to fit the needs of this peculiar common defence against an ever threatening danger.

In our time population pressure and population density as well as the amorphousness of society and its forms of settlement became the dangers, which have urged many states and social groups to plan for the solution of their immediate problems of physical and social structure – and even for their survival. The re-structuring of cities became a condition *sine qua non* for the continued existence of a stable material and cultural standard of life. Therefore the trends of Planning for communities – though certainly not of Planning for concentration of political or economic power – show many similarities, at least within most of the so-called civilized countries. Even our problems of a planned new development of Israel – with all their pecularities of mass immigration and colonization – deviated only in the scale of population-shifting from the problems of other countries, which are planning for their future. In discussing the dangers confronting modern society through the use of the atomic bomb Lewis Mumford sees that the alternative for "the complete disintegration of modern civilization" demands deliberate social renewal. Mumford writes: "To secure even physical survival we must now achieve social renewal." That renewal is the essential trend of our planning of Regional Settlement. It is a long and difficult route that leads to the Planning of our time. If we consider its historic development, we may even come to agree that there exists only one alternative to the large scale social and human disintegration symbolized if not caused by the atomic bomb: the planned Development of regions and regional populations.

I cannot touch here on more than a few prominent points of the historic development of human settlement. We have to imagine a period, when the structure of settlement consisted mainly of rural communities whose farmers cultivated the land cooperatively and in complete harmony with the conditions of nature, especially the preservation of the natural resources of water, soil, vegetation and animals, forming the life cycle and insuring the maintenance of the ecological balance and the permanent fertility of the soil. It appears that the Savah communities of East Asia, of which a few survivors still exist in Indonesia, belong to the greatest accomplishments of this ancient type of human settlement. The word "Savah" does not indicate more than a field, the common property of a village, consisting of terraces and dependent upon a well organized system of irrigation; for the cultivation generally of rice. But around the intricate communal job of the construction and maintenance of the terraces and the irrigation system, and especially the timetable of irrigation, there emerged the whole structure of a village community: its physical, economic and social order, its laws, chronology, religion, family life, its conception of property, values etc.

Around these "simple" and basic problems of cultivation or the preservation of the Savah there grew the culture of a society observing peace with nature as well as with their neighbours, which lasted many thousands of years. The common property of that society was the land, which gave them full subsistence as long as the society preserved the laws of soil and nature. Our knowledge about the primitive communities of European farmers, herdsmen and hunters is generally less complete, but here too it is assumed, that the basis was social balance and the balance between the community and its soil – until the time of the great migration of nations.

An event of the greatest significance for the structure of social settlement took place at that unknown time, when farmers' sons left – or were expelled by social or economic reasons or natural catastrophes from their villages – or at the time, when nomadic tribes invaded the regions of the primordial rural settlement. Events of that kind became the cause for the foundation of non-rural settlements or fortresses, the levying of taxes from the rural population, the origin of the classes of nobility and clergy, the officials, tradesmen, artisans etc. These developments destroyed the basic integrity and stability of the original communal settlements fundamentally; a dynamic social, economic, cultural and physical development ensued in place of the former harmonious stability, which could not be regained till our day. On the contrary, practically all the major changes which occured since that time, made the distance between man and the sources of his livelihood progressively larger and more complicated; likewise they led gradually towards the quantitative enlargement and the qualitative disintegration of social interconnections – the dissolution of communities. A sharp division into a rural, generally subject population and an urban population took place. Agricultural land was converted into an object for exploitation, "big business" and money transactions. The division of labour was enforced in a way contrary to the interests of the rural population. In my opinion it is not just a poetical exaggeration to assert that a kind of a powerful movement of revenge of the disowned and displaced sons against their fathers, the farmers, grew up. (E. Fuhrmann).

Wrench in his book "Reconstruction By Way of the Soil" points to the basic change of values which occurred in the rural regions as a consequence of the invasion of Nomads. The rural population knew of one scale of values only – the fertility of the soil. The condition for the permanent maintenance of this basic value was the preservation of the conditions which the soil demands for its permanent fertility: The rule of return, which includes as intricate questions as fertilization, irrigation methods, afforestation and wild life preservation, intensive cultivation etc. etc. For the wandering and conquering tribes the soil constituted a value of quite another kind. They too depended on the soil for their livelihood, but for them all at once the soil (or: a soil) became the means of getting rich quickly or to ensure their livelihood for a very limited period only. Therefore they came to neglect the intrinsic rules of the soil or to enforce that neglect on the farmers. As a consequence the soil eroded, many regions and many parts of the world were turned into waste land; the decline of Mesopotamia, North America and North China serve only as a few examples. But Wrench furtherly explains that the attitude of the modern businessmen and bankers to the soil is not at all different to that of the Nomads. The economic behaviour of both of them leads to the careless exploitation of the soil, the so called "mining of the soil," which means working the soil as a mine; thus the biological balance in the thin cover of top soil, which is the permanent natural resource of agricultural production, is destroyed in a short time and the top soil itself is washed or blown away by soil erosion. Wrench emphasized what a dangerous illusion it is to make money the cental measure of value in society. Money may be exchanged for an innumerable variety of goods and services – but money is not equivalent to the fundamental natural value given to mankind: the fertility of the soil, which has been and remained till our time the only reliable and the only conceivable basis of life on earth and also human existence. A landowner may quickly become rich by over-exploitation of his land, by robbing his soil of its permanent fertility in the course of a few years, but that fertility on which the permanence of human life depends, will not be restored to the exhausted soil in the course of lying full on for even a hundred years. Wrench demands that the soil be reinstated as the trus measure of value, as the real source of human and social life.

Though the capitalistic epoch marks the most evident changes in the shape of social settlement, it represents basically not more than a consequent continuation or a strong enhancement of the trends of the foregoing epoch.

With the splitting up of social groups and communities goes the technization and specialization of professions. This is the epoch of an immense increase of population, which comes as a consequence of the discovery and careless pillaging of the resources of new continents, the industrial developments, the medical discoveries etc. West-European agriculture is loses much of its importance when cheap supplies arrive in huge quantities from distant countries. A large landless class without property emerges. Cities grow quickly and become fortresses of wealth. Industrialization progresses, mechanization overwhelms social traditions and injures the convenience of life. Contrasts and hostilities between classes and nations as well as between individuals grow. Living expenses increase with the decrease of the actual standard of living of the masses. The technical possibilities of society become wider, but no new values of social life emerge and the social structure disintegrates; technization becomes an independent powerful factor. I am speaking about those features of the capitalistic epoch, which have been described so convincingly in the books of Lewis Mumford amongst others.

The structure of settlement in one of the states of the U.S.A. may serve as an illustration. In Europe the picture of the typical forms of settlement created by modern civilization is mostly indistinct because of the long historic development of villages and cities, the deeply rooted traditions of building and urban and rural Planning and regional life, keeping its ground. But in many parts of the U.S.A. this civilization built its proper form of settlement on *"tabula rasa."* Here it becomes evident that modern civilization did not at all intend to settle permanently in the regions of "the good earth," but to keep on moving in the search for ever greater riches. Its settlements show all the features of temporariness. The farm houses are constructed as feeble wooden shanties; the low standard of building is striking. It is the tendency of many farmers to get rich as quickly as possible, so as to acquire a comfortable house in the exclusive suburb or a large city.

The sight of some of these small towns is even more depressing. You do not find their many solidly constructed houses, but a type of a huge camp divided into rectangular blocks and small rectangular plots, where everybody according to his taste and his means errected his temporary dwelling. Though living conditions in these houses are often very poor, care was taken to give them an extremely pretentious street elevation. But on the other hand the inhabitants of these houses – even those with a comparatively low income – are the owners of a surprisingly rich and complete outfit of movable apparatus: a big car or two, a radio for every member of the family, a washing and drying machine, a television set (to have one becomes a matter of social position) etc. The centre of this small American city with its typical Main-

Street is a limited quarter of very high office buildings (10 - 20 storeys). Here the cafeterias glitter in rich brightness, day and night advertisements dominate all the views, gasoline stations compete with their dazzling colours – it seems, that all the services are estblished for attracting passing tourists only, but not for the needs of a permanently settled, balanced population.

Let me qualify this description, for it would be unjust to assert that this "style" of city is typical for the whole of the U.S.A. In many parts of the States settlement is of quite another character. It all depends on the origin, the tendencies, and the period of settlement. Nevertheless it seems that these examples of American settlement in the capitalistic epoch are rather instructive. They indicate that society returned to the attitude of the migrating tribes. Society tends again to "mine and move." Its "stability" rests ". . . on its capacity for promoting change . . . its safety . . . upon its progressive tendency to revolutionize the means of production, promote new shifts in population and take advantage of the speculative disorder . . .".

The basis of modern Sociology and Planning is the realization of the dangers for the existence of society latent in this development. At the same time they represent one of the reactions to it. It is their basic attitude, that the stability of settlement of communities on the land, in villages or cities, ensures peaceful social development and enables man to further his greatest possibilities; neither the forced migrations of population nor the spontaneous pursuit of the "miners" of all kinds after quick gains and profits, are ways which give any hope for the social present and future. The resettlement of communities, geographically and economically – that means the new regionalism – therefore has become one of the important positive aims in our day. Now neither sociological Planning of land use are able to achieve their social objectives directly, still less can the planner enrich our purpose in life. In reality the Planners cannot do more than try to prevent the blocking up of "the road to survival," to render possible the renewed integration of communities in town and country. The Planners' analysis of our reality indicates that man's settlement in the cities of the last century bears the marks of a temporary occupation. In fact these cities are not more than transit stations of a migrating population – an unsteady and often erratic economic life, exposed time and again to rapid changes. Thus in many parts of the United States the location of cities does not appear to represent a durable location, and there is even reason to believe that the final colonization of this immense country has not yet begun. West Europe, too, needs reconstruction to a great extent as a consequence of the change in its relations with overseas countries, and East Europe because of the loss of markets for its industrial production. These countries cannot any more build their existence on the exploitation of the resources of distant countries. They have to start a new to work on their own land and to develop their own natural resources. In fact Planning in the Netherlands and England represents a first step towards this new reconstruction and stabilization.

In one of his recent articles Prof. E. W. Hofstee, the Dutch sociologist, discusses the three types of basic personality or the three forms of "life style" of agricultural population as proposed by the American sociologist C. Zimmermann: The Homeric type (his central interest – the family); the Hesiodic type (central interest – the farm and the land) and the Aristophanic type (whose main interests are monetary profits). Zimmermann argues that at periods when the Aristophanic type becomes the prevailing type, differences between rural and urban life, in values and habits of life as well as in types of personality, dwindle and there remains the difference of profession and dwelling place only. Hofstee agrees with this analysis, but he argues and verifies that the Aristophanic (markets) type is no type of independent basic personality at all, "but a *temporary transitional type* only, appearing just because of the *lack* of a distinct style of life...". This type appears when man does not feet himself tied any more to society and its mode of life, so that his egoistic instincts become pre-eminent.

I believe that this criticism of Hofstee points to an essential Planning attitude. We have to admit that the Aristophanic type addicted to profitmaking without any consideration of soil, family, and society is still the ruling type in many respects. But we have to analyze this type as an appearance of a transition period only. In our Planning we assume, consciously or subconsciously, that the communities that will finally inhabit the new planned villages and cities, will be composed of stable human elements, essentially interested in the earth, their village or city, family, creation of stable values and their region. We ought to admit that our assumption it not realistic – at least not for our immediate present, but then it would not be worth while at all to go on planning without that "unrealistic" assumption.

It seems that many of the difficulties we meet in our daily work – especially in negotiations with the "realists" of many kinds – have their cause in this "unrealistic" attitude of ours. It is true, that we base our proposals on scientific studies: clasification of soil, soil conservation, water conditions, needs of agricultural economy, industry, transportation, open spaces and recreational areas needed for every inhabitant, and so forth. But at the same time the preparation of a regional or local scheme – the coordination of all the technical details – is not exact science, since it is done on the basis of an assumption only, the assumption of the preservation of the laws of nature by man and stable communities willing to settle permanently on their land. Such assumptions are basic to the Planning policy. Our basic schemes are land use schemes; apparently their preparation is a technical-scientific job, but in fact it is the implicit presupposition of the Planners who work on these schemes that man and society will eventually decide to use the land not for short term profits, but according to its natural laws and possibilities, and for the sake of preserving forever that central value. This means an assumption of a decisive change in current methods of social development.

It may appear, that this Planning attitude is related to agricultural settlement only. But in fact it extends likewise to urban settlement. In our period a number of cities grew into huge conglomerations. You may say, that the inhabitants of these metropolitan cities consists of the surplus in varying degrees dependent upon exchange with its "super-region," which may be of the fast-increasing population. What are the tasks of Planning with regard to this section of the population, which apparently found its fixed site of settlement? Is it more than to ease the troubles of density, an effort comparable to the task of the traffic policeman? Indeed this is the purpose of local planning in many countries. Not only in most parts of the U.S.A., but also in many European countries Planning confines itself to proposals for the improvement of conditions in several densely congested big cities; as to the Planners – they find a field of activity in these cities just because in these countries the bigcities are the only institutions rich enough to bear the cost of Planning. Although this kind of Planning is certainly very important for the improvement of living conditions, reduction of congestion, and dangers to life and limb of motorized traffic, the actual achievement of these aims generally succeeds to a very limited degree only. Even the new neighbourhoods planned on the outskirts of these big cities cannot sufficiently get rid of the direct and indirect influence of the conglomerations to which they belong. This kind of Planning does not go beyond the frame of a defensive improvement, which aims at the artificial maintenance of an essentially infirm way of development.

The whole weight of social criticism of current Planning becomes clear only on the regional level. In the frame of regional research and Planning, it becomes obvious that our big city is not at all a permanent way of settlement, but the temporary concentration or fortress of a nomadic population, alienated from stable economic, social and cultural values; the residents did not really take roots in the big city, but ... "are to a large extent strangers ... seeking sanctuary or fulfillment ..." (E. B. White) and these "sanctuaries" survive with regard to their economic, social and cultural existence, on a speculative basis exploiting vast areas or even continents, but cut off from their immediate natural surroundings. In their search for the factors favoring the consolidation of human settlement in our time Planners with an ecological background look upon country and city as on parts belonging to a larger unit – the region. It may be a small or a large region, giving subsistence to one small or big city, or a number of cities, in accordance with the natural wealth and the variety of its resources. A region may be the part of a country, a whole country or a continental unit.

Indeed it is an essential part of the regional idea that regions join together to form superregions to even larger regions, and so on towards some intercontinental or global super-unit. But intrinsically regionalism means the formation of social-geographic nuclei, the regional communities, capable of living largely by mutual exchange within an easily conceivable regional unit. The region aims to develop its own available natural and human resources and so to educate its population in peaceful productivity and the understanding of the endeavours of other regional communities and other people. The region is the environment into which settled man and society organically belong.

The stability of a city's existence will be based on a region with a balanced resource development, and on the establishment of living ties between the city and this region. The physical result will be that the plan of a new city designed on the regional basis will represent neither a closed defensive or aggressive fortress, nor an accumulation of huge blocks of buildings turning their back to the countryside, but a concentrated settlement opening itself by green wedges to all parts of its environment; it is just the organic connection with this region that ensures the stability of these cities. The village too depends in our time on supplies coming from town, as it is no longer selfsufficient; it relies on the urban services, markets, industry, artisans. Regional Planning thus considers the city people and their rural partners as one super-community with common interests in all the details of development of social life – an ecologically balanced unit or complex of men. For the Regional Planner the desirable solutions become feasible only when city and village are considered to live in a kind of symbiosis: this presumes essential differences in the character of urban and rural life, but points to an economic and social equality of rights and mutual help. It is true, that we have to take into consideration that in most countries the urban population will exceed the rural population in numbers because of the simple fact that one agricultural family will produce the food of four to five – or more – urban families. But with regard to the complete region, town and country will become units of equal value balancing each other in their economic-social weight, if both of them tend towards long range economic security and stability of settlement. The isolated city, carelessly neglecting the development of mutual relations with its proper region, endangers its agricultural and natural environment: its economic basis, because the city may obtain its food supplies from distant countries according to momentary opportunities; its land, because of its reckless and disproportional expansion. The isolated big city turns agriculture into a speculative business, producing in mono-cultures for unsafe markets; it endangers the style of life as well as the existence of the rural communities. But it also endangers its own safety, enhancing its own economic insecurity with every change in national or international development. It is only in coordination with its regional framework that the city finds its final location, its natural size and its stable tasks, so that the ties with its sphere of influence become distinct, direct, and close ties. Living conditions in this regional city become healthy, because the reasons for conglomerations which are out of proportion for man and society are minimized with the spread of regional development. Thus it becomes possible for man to survey the factors which nourish him, which create his living conditions and the character of his society. In the regional enclosure the urban citizen too may take roots on his urban land.

All this does not mean that the economic life of a region must necessarily represents a closed cycle of communities producing only supplies for the regional population. In most countries industry and agriculture will have to produce continually for national and international markets as well, and part of their food and other products will have to be supplied from far off regions. Consequently realistic Planners have to take into acount that Regional Planning will never constitute a system offering solution for all the Planning problems. But here the question of the predominating trends becomes decisive. These Planning trends point towards the enhancement of cooperation and mutual aid in the regions which will characterize and form the basis of all social and cultural activity. It is the tendency of Planning to create the physical conditions for the renewal and the stabilization of community settlement on the rural as well as the urban land. "Not mine and move, but stay and cultivate are the watchwords of the new order." (L. Mumford).

In the foregoing words I intended only to stress the new Regionalism as a necessary and significant reaction to the state of civilized society today. But in fact economic and social life, as well as the physical aspect of city and country, are influenced in many parts of the world, especially in Europe, by ancient traditions and geographical, economic, and political situations. The Regionalism of former periods has not yet disappeared everywhere. Thus in the Netherlands the traditional Regionalism serves as an excellent preparation for the development of the renewed Regionalism, though its economic and social contents do not represent an adaptable basis for the development of modern industry and agricultural reform. In Israel our situation is different. During my stay in the Netherlands I was told by many Planners that they envied us, as conditions for planning seemed to be ideal in Israel with the larger part of the land in public ownership and the new immigrants settling and developing the country according to plan. In this respect possibilities in the Netherlands are of course much more limited; but I always had this answer, that I found Planning favoured in the Netherlands because of one important reason: the availability of the most important factor for planning, which is the reliance upon the human element in firm and educated people prepared by historic experience, and capable of developing a renewed regional life on their common ground, in the villages and cities.

Though the quality of the human element represents for us an "unknown quantity," we in Israel may profit greatly from the experience in regionalism of other countries as long as we bear in mind the differences. I want briefly to give here a few illustrations on two important questions of regional Planning which are bothering us considerably: the question of the integration of agricultural and urban social-economic life, and the question of the size of new regional cities. The structure of settlement in the northwestern part of the Netherlands (physical and structural conditions in the Netherlands resemble in some important points conditions in Israel) is composed of four main elements. The first one is the agricultural farms, each built on the proper agricultural plot of the farmer. This means a wide dispersal. The second one (the so called "A Kernen") is the village, the smallest concentration of population of 500 - 3000 inhabitants. It is in only a few cases the dwelling place of farmers; it mainly serves as the residence of agricultural labour, artisans, specialists and the population rendering the immediate local services to the farmers. These villages are dispersed at a distance of approximately 7 km from each other. "B-Kernen" are small rural towns of 5000 - 15000 inhabitants approximately at a distance of ca. 20 km from each other, concentrating trade and rendering services of higher degree for a whole subregion, including also some agricultural industries. It is rather important for us to learn from the excellent Dutch sociographic studies, that these "B" centres have been losing importance in the course of the last decades. The reason for this decline may be found in the development of modern means of transportation, the variety of standardized supplies, the modern system of distribution of products, cooperative societies etc., which make the farmers and villagers less dependent upon the neighbouring "B" centres. Nowadays, the villages shops are able to offer the farmers a large part of those manifold supplies which in former times counld only be found in the "B"-centres. Thus the social-economic position of the villages ("A"-centres) has grown. But at the same time there has taken place a healthy development of the "C"-centres – the medium size cities of more than 25,000 inhabitants situated at distances of about 30 km from each other. These cities attract more and more industries and service trades, which in former times were located in the really big cities on one hand and in the "B"-centres. Especially, industry develops well in the "C"-centres, which offer favourable conditions for its technical and commercial needs, as well as big advantages from the point of view of social welfare and healthy community development.

One of the important consequences emerging out of the Dutch socio-graphic research, which is of great significance for us too, is the clear conception that city, village, and farm form in fact an indivisible economic-social and physical unit, and that its parts cannot be treated separately, if the efficiency of Planning is to be preserved. But the condition for the achievement of the regional unit is, that the parts are sound, complete and balanced in their mutual relations. This condition we have to bear well in mind, especially in respect to the size of the regional urban settlements and the occupation of their population. It would be dangerous to try simply to "transfer" the traditional regional structure of some parts of Western Europe to Israel, and every mistake we comit here may upset the entirety of a whole region. Besides the change in regional structure coming as a consequence of modern industrial and transport developments, which cannot and should not be reversed, we have to reckon with the cooperative character of agriculture in Israel, and the essential fact that the inhabitants of the cooperative settlements represent one of the strongest social and cultural factors in the structure of its population. I believe that the fact that we cannot even imagine the exploiting of our cooperative farmers by the inhabitants of a regional city, proves to be an advantage. On the contrary, the danger exists that our agriculture will hesitate to utilize the services which the new regional cities are able to render; thus the regional development may be delayed; proletarian cities without any permanent social-economic ties with their environment may grow up in the middle of well developed and organized agriculture regions.

This danger has to be taken seriously into account in our Planning. It sets limits to decentralization and dispersal of urban settlements. You are undoubtedly aware of our proposal for the division of the country into Planning regions. Today we have to tackle the problem of distribution of urban population within the Planning regions. It is true that the planning of the economic destination and the size of the new cities depends in every case upon the special conditions found on the site. But the economic, social and cultural attractive power of the regional city for the agricultural population as well as for industry will increase, if the urban development of the regions will concentrate on "C"-centres (cities above 25000 inhabitants) with a variety of administrative, economic and cultural functions. It appears that many more considerations confirm this hypothesis. To build up a regional balance suitable to our conditions, our period, and our wish for a decent standard of life for all inhabitants of a region, asks for the organizational and economic strengthening of urban as well as rural settlements within the regions.

In every country that depends for the achievement or maintenance of its balance of payment upon the further development of industries, this problem is of special importance. Wherever industry has to be "artificially" developed, there the question of location of industry and the degree of its concentration will deeply influence the state of regional balance. In this respect the Planning policy of some Catholic provinces I visited in the Netherlands is of great interest in so far as it may serve as a warning. Apparently they subsequently accepted the theory of decentralization. Thus the development scheme for one of these provinces shows its division into 46 regions with 46 industrial centres of population of mostly not more than 1000–20000 inhabitants. The distances between these centres are 8–10 km only. This scheme has its tendencies too – the tendency to keep the fast-increasing local population in its own surroundings, free from the "seductive influences" of even a medium sized urban concentration, and to maintain in this way the very elaborate system for the rule of the Catholic Church. In fact the consequence of this fear of "social and moral disintegration" may be the cause of removing the favourable opportunity of some places to grow into cities of a reasonable social and cultural level. The excessive dispersal of urban populations into numerous small centres may lead at the same time to the estrangement of rural and urban populations, as these small centres are too weak to fulfill any regional functions, too one-sided in their occupational structure, and economically and socially too unattractive for the agricultural population. Agriculture in our days does not demand such a dense network of secondary service centres. Of course the danger that these small centres will develop into poor proletarian agglomerations and upset the regional units is obvious to the majority of Dutch Planners.

In large parts of Israel, Regional Planning today is realized in the actual construction of cities, villages, roads, installations, and plantations. This important activity should not foster in us any illusions. I wish to emphasize again, that the tasks and possibilities of Regional Planning with regard to the actual development of social regional life are limited. Planning cannot

achieve more for the time being than to point to possible ways of social integration and to realize physically several important preconditions for permanent, stable urban and rural settlement. But there are no prospects of
success for regional development, if the realization of regional settlement is
not accompanied by good will, understanding and help on the side of the
farmers and their organizations, and by the instructors and teachers who will
assist the new inhabitants in the stage of transition and acclimatization. For
the period of transition we need regional development societies, for whom
the development of regional economy and the mutual relations of Town
and Country represent the most important task; we need not only professional instruction but also the education of the new inhabitants towards the
formation of regional communities. As a matter of fact we need, for the development of the regional cities, pioneer settlement organizations of no less
strength than our agricultural organizations. We should never speak about
the achievement of the realization of our regional schemes so long as the
social attitude of Planning does not find its continuation in a broad public
activity, aiming at even broader cultural goals.

VI

THE RELATIONSHIP BETWEEN LANDSCAPE PLANNING AND REGIONAL PLANNING

For hundreds of years the function of Landscape Planning has been the design of landscape artifacts such as parks, gardens, waterworks for the leisure class. To spend part of one's life in a landscape where man had succeeded in over-ruling nature was considered a luxury which could be attained only by the richest families. To live close to nature, adapted to ecosystems, to topography and to the seasonal changes of climate was the regrettable fate of the poor peasants, shepherds and woodsmen, and even of townspeople in times of distress.

In many respects this situation has been reversed with the growth of urbanization. The majority of the poor or the middle classes have to spend their lives in a man-made town and landscape of living, working, recreation and communication, and only the richest classes – or outsiders – can permit themselves a "return to nature" in the diminishing areas of landscape where the control of change is still mainly biotic. These landscapes can be kept in their natural condition only as long as the number of their visitors does not exceed certain narrow limits. The dilemma of many landscape planners is that the natural landscape – at least in the more densely populated countries – cannot be preserved for the enjoyment of a considerable part of a population; on the contrary, for the sake of preserving nature such pressures of population must be resisted, and obviously there is a limit to maintaining such resistance.

The rise of living standards and the increase of leisure time for the masses of people in our time has led to increased requirements for larger and better areas of accessible landscape. Large sections of population have turned to partly nomadic ways of life and their dependence on landscape as a habitational and recreational environment has vastly increased. In this situation, the success of preservation of natural landscapes depends paradoxically on the planned expansion of man-modified landscapes over a part of the natural and all over the rural landscape, as well as over the urban peripheries. We must enhance our awareness of the fact that a man-modified landscape and "nature" (biotic or historic) will be the environment of many creative and re-creative phases of human life. The question remains if the man-modified landscape will be the chance result of land-use controversies or exhibit a deliberately designed and controlled environment.

The contemporary landscape has become the battlefield of contradictory economic, social and biotic pressure and processes. Such contradictions lead to a general landscape deterioration. This fact has raised a demand which is practical and ethical at the same time. Having become the ecological dominant over most of the surface of the earth, man cannot rely any more upon nature to counter-balance his impact on environment; he has to control landscape and resources actively in a planned way. We have to renew our awareness of this responsibility and to relate it to the increase of our knowledge, power and scope of our problems. In the present situation, the problem of the landscape planner is not any more the preference for either a man-made or a natural landscape; but the establishment of social and biological controls which ensure the survival, usefulness, and value of landscape as a life-resource.

The primary instrument of control of the man-influenced environment, developed and applied in a few countries only, is Regional Planning. In the best cases Regional Planning deals with the economic, social and physical-environmental aspects of change. The biotic aspects are still neglected and as a consequence the basis of designing and controlling the regional environment is defective. To become fully effective, Landscape Planning has to form part of Regional Planning; but in this framework Landscape Planning will contribute decisively to the comprehensiveness of Planning by introducing several new aspects.

For the design and control of changes of contemporary landscapes, as well as for the defence of whatever natural or historic landscapes still exist, we need clear definitions of the functions of a landscape. Such definitions should touch on social and biotic, economic and ecological, rational and emotional aspects. Contemporary landscape functions needing coordination, integration or separation have been defined by the Committee on Landscape Planning of IUCN as follows:

Sustained intensive production of food and raw-materials.

Habitability and attractivity of the environment of the permanent rural population.

Recreation for urban and other temporary or mobile population.

Manifestation of biological and cultural continuity as well as of human self-expression.

It might be argued that an all-pervading phenomenon such as Landscape is, cannot be exhaustively realized by ascribing to it several "functions." Indeed, I believe it will take some time before we shall be able to grasp and to communicate the full significance of Landscape and its effect upon the maintenance and quality of human life. In the present crisis of landscape use, however, an attempt at rationalization, even if incomplete, has become a necessary measure of standing up for the planning and the carrying through of landscape development controls. Even such a provisional clarification of landscape functions has the merit of opening a way for the integration of Landscape Planning and Regional Planning. The preservation of existing landscape values as well as the creation of new values depends largely upon an attitude which recognizes right from the start of the Planning process: The landscape aspects of any Regional Development Scheme.

The aspects of human settlement, mobility, work and leisure in any Landscape Scheme.

In the following paragraphs brief descriptions are given of three cases of landscape problems and Planning proposals which can be properly solved only by integrated Landscape and Regional Planning.

The first case is that of the East Flevoland Polder in Holland, the country which has made the most significant progress in planning its cultural landscape. The problem of this Polder is one of rural-urban integration in a new cultural landscape.

The second case is the Island of Crete where Regional and Landscape Planning may still come in time to avoid the confusion of uncontrolled urbanization and rural decline.

Finally the Lakhish Region in Israel will be mentioned, where a generally successful implementation of a rural settlement scheme will have to be complemented by Landscape Planning on a regional scale.

The East Flevoland Polder

The East Flevoland Polder is the most recent of the Dutch reclamations from the sea. One hundred and twenty-five thousand acres of the most fertile land have been added to the area of the Netherlands and are being transformed into a living region by the planning of the combined economic, social environmental and technical aspects of development and life. The broad experience of the Dutch in the planned development of new regions of settlement comes to good use in this new polder.

The northern limit of the area is separated by a narrow stretch of water for the North-East Polder. The latter had been planned and has become by now an exclusively rural region, wherein all economic enterprise is based on agriculture and the services required by a well-to-do rural population. The landscape, villages, and a town of 9,000 inhabitants, Emmeloord, have been laid out to suit the needs of land cultivation, drainage, efficient transportation and communications, and of the viability of rural settlements "islanded" in a sea of open food production areas.

The south-western dike of the new polder is only 30 miles from Amsterdam and the metropolitan and industrial "Randstad" region of the Netherlands. This is a horse-shoe shaped series of towns, comprising the two great ports, the capital and other centres of the country. Over four million inhabitants live in this region. In this dynamic European centre land has become the scarcest commodity, for habitation, cultivation, industry, commerce and recreation are in fierce competition for land use.

More than half the East Flevoland Polder is already under cultivation, while the rest is still marshland. Some new villages, a small town and many farmsteads have been built. On the basis of a general Landscape Plan prepared by the Government's Landscape Planning Department, the southeastern coast of the polder, facing the old land, has been converted into an attractive recreational area – a completely artificial creation of freely curving beaches and islands – built of sand pumped up from the bottom of the sea. The area has been planted with whole plant-societies as a partly open and partly enclosed landscape. In fact this landscape is indistinguishable from any "natural" lake beach, except for the height of the trees. In summer this area is crowded, or rather overcrowded, by vacationers, trailers, tents, kiosks, and boats. Nearby forests have been planted to expand the recreational area in future. Observing this area, the artificial introduction of "natural" lakes and landscape does not appear as a romantic reaction to urbanization but as a legitimate means of design.

However, the "inland" of the new polder, as far it has been developed, is a flat and desolately monotonous landscape, surpassing in this respect the North-East Polder. The landscape of the latter has been criticised for its rigidity and inhuman scale, but in this respect the new polder goes to extremes. This is a result of the increase of open field units and farm plots, and the increased adaptation of land planning to mechanized cultivation methods.

These visual contrasts in the East Flevoland Polder are an indication of a conflict over landscape formation involving some essential questions of contemporary use of land and landscape. In the "affluent society," the rationalization, commercialization and mechanization of agriculture are leading to "scale-enlargement" – i.e. an increase in the areas of influence and the distance between farms and villages – and to an absolute decrease in the number of farmers and farmhands needed to work the land and provide services. With the expansion of monocultures and the population decrease the rural landscape loses its traditional function of a useful and habitable environment, and becomes a nondescript food factory. The result of these changes may be called a "Rural Exodus," the word "exodus" indicating the loss of population, intensity of rural life and landscape values.

These tendencies must be considered side by side with other factors exerting quite a different influence on this polder landscape. The impact of these factors is at present only partly visible, but it will be felt much more strongly in the future.

The first is a decision in principle to integrate the economy, society and the transportation system of this new region with the Netherlands as a whole. This trend is exemplified by the plan for a national "development axis" crossing the East Flevoland Polder and connecting the local pattern of rural settlement with the metropolitan region and the countryside pattern of settlement.

The second factor is a plan to establish a town of 100,000 inhabitants (Lelystad) in the region. Its proposed functions include harbor and industrial development and are of national rather than regional importance. The establishment of this town will raise the level of all regional services and influence the "cultural climate" of the population of all the polders. The interest of an urban population in building and living in the mindst of a cultural landscape – and not of a food factory – may exert a beneficial influence on the future face of the polder. Landscape will be planned for the benefit of both urban and rural population.

The third factor is the increasing need of recreation land, now felt in the Netherlands to such an extent that it is considered "economic" to reclaim at tremendous cost new fertile polder land solely for this purpose. This consideration has for the first time brought about practical results on the south-eastern coast of this new polder, and all signs indicate that this is only a beginning. What mobility and the lack of recreational areas lead to in Holland is a renewed urban interest in landscape values, both in specialized recreational zones and in the rural landscape in general.

The fourth factor is the demand for a more habitable, accessible, and nationally integrated rural environment and more attractive rural amenities. The demand comes from prosperous farming communities and is strongly supported by agricultural and forestry authorities in Holland. This demand may be considered an indirect consequence of the increasing influence of "urban values" among the rural population. The latter, particularly in the new development areas, cannot be kept on the traditional cultural and environmental rural level; for them the rural and the urban levels must be comparable; which means they must be complementarily related to each other. The rural environment's own resources of habitability must be developed.

The fifth factor is the Dutch Government's active interest in Landscape Planning which has become an integral part of the land development program. The work done now in all parts of the Netherlands by the Government's Landscape Planning Department, under the direction of Roelof Bentham, can serve as a model for many countries. Ecological considerations have been taken up in the Landscape Planning programs. To balance the impact of mechanization and commercialization of agriculture on land and landscape, suitable plant associations are established along field borders, roads, near villages and coasts; they will be an essential feature of the future polder landscape.

In the East Flevoland Polder the problems to be faced become most evident. Here the rural exodus is clearly countered by an urban influx of population and culture. Most of this population will consist of temporary visitors, but with the decline of habitability in some metropolitan areas, the presence of permanent urban inhabitants in rural regions must also be envisaged. Perhaps some groups of highly intensified "microfarms" must be laid out to receive this movement of return to the land. A basis is therefore being laid for the development of diversified rural-urban regions serving a multitude of environmental purposes and desires.

It is doubtful whether the economic, technological and other functional factors leading to the rural exodus and the urban influx can be considered as sufficient determinants for the formation of the future landscape of the new polder. Many Dutch planners, particularly the rural sociologists Professor Hofstee and Dr. Constandse, have become aware that affluence and technological development have increased the range of alternative solutions for problems of settlement structure. Economic considerations of the structure of a new settlement region do not by themselves establish any single solution as best. What is needed, therefore, is the formulation of environmental desiderata and their expression in the planning and design of regions. In fact, this measure of freedom in decision – or this need "to make up one's mind" on the environment desirable for men, on *what ought to be* the use of the earth, was postulated many years ago by Patrick Geddes. He considered the expression of the ideals of an environment to be a vital ingredient of all practical planning and a necessary counterweight to the conditions and situations determining, as it were, a plan. What the Dutch planners have done so far in this respect is the introduction of an unusual measure of flexibility in the plans for the East Flevoland Polder. Though no planning body has made more thorough studies of the various functional determinants of the region's development, many possibilities for change and new adaptation have been left open for the future.

In the East Flevoland Polder the problems to be faced by a Landscape Planning policy become most evident. Here the rural exodus is clearly countered by an urban influx of population and culture. Most of this population will consist of temporary visitors, but with the decline of habitability in some metropolitan areas, the presence of permanent urban inhabitants in rural regions must also be envisaged. Perhaps some groups of highly intensified "micro-farms" should be laid out to receive this movement of return to the land. A basis is therefore being laid for the development of diversified rural-urban regions serving a multitude of environmental purposes and desires.

The location of the East Flevoland Polder in a transitional area between a rural development region and a metropolitan region intensifies the problem posed by the formation of this landscape; it might be considered for this reason as a project of crucial general importance. As a result of the "rural exodus" and the "urban influx" the regional balance made up of rural and urban elements, values and populations is undergoing a process of change. Urban-rural integration is taking on new forms and the use of land and landscape in rural areas is becoming ever more a regional community interest.

Landscape Planning on Crete

In 1964 a comprehensive Development Plan for Crete was prepared by a team of Greek – especially Cretan – and Israeli planners and specialists. In the section dealing with the physical aspects of Regional Development, a general concept of a new structure of settlement and landscape, to be achieved in several stages is laid down in the form of a "Sketch plan." This plan is characterized by a "built-in flexibility." Six definite targets, however, are considered as basic for the achievement of a consolidated and meaningful settlement structure. One of these main targets is the reconstruction, by soil conservation measures and afforestation, of the uninhabited and nearly uninhabitable mountain landscape of Crete, and the creation of a continuous landscape belt of varying width, all along the ridges.

The "Sketch Plan" postulates that "structural change in economy, environment and society has begun in Crete, and must be further accelerated by Planning. At the same time, the preservation of cultural and environmental values in the process of change should be one of the major objectives of the planned development of Crete." The proposal for Landscape Development is conceived as furthering the achievement of this double aim.

Large areas of the mountains of Crete, are uninhabitable or very sparsely populated. Part of these areas serve as marginal pasture land, but nearly all of them are treeless and bearing the marks of severe soil erosion. This situation is not the result of natural conditions, for ancient Crete was an island of woods which were later destroyed by wars and mismanagement. For a number of reasons the restoration of these woods is of importance for the new regional development of Crete.

In the agricultural development plans for Crete consideration is given to the afforestation of areas which are unsuitable for permanent pasture. This proposal should find its place in the larger framework of a Regional Landscape Plan considering a variety of human, biotic and physical aspects. The reconstruction of landscape is to form an integral part of the comprehensive Regional Development Plan.

The extent of this project of Landscape Planning is presented on a map in which the uninhabited central and coastal mountain areas of Crete are defined and interconnected by strips of landscape wherever the Cretan main-watershedline passes a valley. In this way it is proposed to create a "landscape system." Landscape Planning in these areas will serve a number of purposes:

The prevention of damage by floods and soil erosion to agricultural land, by planting ecologically suitable plant associations; the protection of dams and improvement of the water supply for irrigation; wood production in a part of the afforested areas; improvement of micro-climatic conditions; the cultural need for re-creating the indigenous mountain landscape of Crete; and finally it will result in the creation of a hospitable and accessible mountain landscape for the sake of general habitability of the regional environment and for recreation.

To serve the multiple use of landscape, detailed Research and Planning will be applied, defining the treatment to be given to each particular section of the landscape. Selective afforestation and soil conservation measures will be proposed accordingly. The mere planting of monocultures of eucalyptus or pine trees on otherwise "useless" land should be considered as a dubious measure for using the mountainous land of Crete.

The over-all purpose is to create a continuous belt of varying width of protected mountain landscape, including also the highest mountain reaches beyond the tree limit. This belt will extend from the western to the eastern coast of Crete, with "branches" connecting some coastal mountain areas to the main system. On the ridges, or nearby, a continuous trail will be laid out after detailed investigations, leading over the Levka mountains, Mount Ida and Mount Lassiti. This proposal is based upon the idea of Benton Mac-Kaye, the originator of the Appalachian Trail in the United States which passes from New-England to Georgia through a continuous protected mountain landscape. Planned forty years ago, the Appalachian Trail is today proving a tremendous asset to life and landscape in the Eastern United States. The general topography of Crete is particularly suited for the realization of such a "macro"-landscape scheme.

The whole "backbone" of the Island of Crete should be turned into a kind of National Park. It would be touched upon by motor cars, wherever North-South roads intersect the landscape system. For the rest, it would be accessible only to foot travellers, pack horses and mules. Camping grounds and accommodation for tourists in some nearby villages would be established along the "Cretan Trail" but, learning from the unfortunate experience of some European countries, no private development of villas etc. should be permitted in this National Park. In these large areas, a unique kind of recreation and tourism could be developed.

The project, which has been laid down for the time being only along general lines, aims at fully integrating the needs of Regional Consolidation withTourist Development.

The Lakhish Region in Israel

The newly developed region of Lakhish, in the south of Israel, has no distinct natural boundaries. Of the 70,000 ha. the western part in the coastal plain which comprises about 40% of the total area, consists of potentially fertile land. The eastern part, approaching the foot of the Judean Hills and the Jordan Frontier, is heavily eroded hilly ground, full of the vestiges of an ancient culture, and mainly suited for pasture.

Because of the scarcity of local water resources, this area was under-developed and under-populated until a few years ago. Its development began at the end of 1954 according to a comprehensive regional settlement plan. Execution was coordinated with the laying of a 60 km. long water pipe-line from the sources of the river Yarkon to the South of the State. This was the technical pre-condition for the intensive cultivation and modernization of the neglected district.

The most important objectives of this regional settlement plan were the increase of agricultural production, the integration of a destitute population of new immigrants into the life of a regional rural community, and the intensive settlement of an important part of the State. The Planning Target for the Lakhish Region, which was originally a population of 40,000 with 50% farmers, is now 60,000, 33% of whom will be farmers, and rest will work in services and industries to be located mainly in a new regional centre, the town of Kiryath Gat.

The planning of the social and economic services of the villages is one of the most interesting features of the Lakhish scheme. Wherever villages can be placed close together – and this is possible only in areas of intensive irrigated farming – they are grouped in clusters of 5 - 6 villages around a service village, the Rural Community Centre. In this way small sub-regions are established, and a regional structure, constituting a new pattern of settlement, is created. Kiryath Gat, the centre, is surrounded on one side by a semi-circle of four sub-regional "clusters," at an average distance of 10 km., and on the other side, which is a hilly and less fertile area, by a number of more self-contained villages or "Kibbutzim" dispersed at a distance of about 10 - 15 km. from each other.

The old landscape of the Lakhish area, characterized by cactus hedges, densely built mud-brick villages, sheep-folds, picturesque wells and lonely clumps of trees, has disappeared completely as a consequence of the new development. The newly emerging pattern of landscape may be considered as the result of the re-distribution of the land according to modern concepts of rural regional development, of mechanized cultivation methods, of irrigation farming, and the building of standarized, comparatively extended villages, consisting in the first stage of small dispersed houses. Mechanization of land development and new construction on a large scale are determining the character of the present landscape. It may be said that the landscape of the Lakhish area has lost many of its esthetic values. Its esthetic poverty, however, is closely related to certain functional deficiencies in the regional pattern.

A Regional Landscape Development Plan which, it is hoped, will be realized in the near future will correlate the functional and the esthetic aspects of the developing Lakhish region. The approach to this project and its goals are determined by the consideration of the needs of both the region and the country. In the course of the past years great progress has been made with the economic and social development of the Lakhish region. These successes make further action for the consolidation of permanent settlement particularly important. On the other hand, the urbanization of the region of Tel-Aviv-Ashdod has rapidly expanded during the same period, and some of its reflections are felt also in the Lakhish region. A new development stage in this area will accordingly have to serve a double goal: the enhancement of habitability and attractiveness of the region as a place of permanent settlement, and the absorption of a part of the recreational movement "radiating" increasingly from the urbanized areas.

In the Lakhish region the proper home of the population is considered to be the region rather than the individual village. The regional space, therefore, will be a much frequented environment, and it must be designed and treated as carefully as any urban space. At the moment, in the intensively settled part of the region, the landscape has merely the character of a food-factory, lacking trees, shadow, variegation and human scale, and the villages have the character of inhabited islands in the open landscape. It is in the regional landscape as a whole, that the community must find its environmental self-expression.

Any regional structure is deficient, if the roads and communications are "uncivilized" and uninviting and if the approach to and the form of the regional foci are unattractive. Such requirements gain particular importance in an area of difficult climate conditions such as those of the Lakhish region. The creation of comfort all through the structure of the regional landscape and the creation of environmental values will enliven regional inter-action and enhance the emotional connection between the settlers and their region. In this way it may contribute decisively to the consolidation of settlement.

Some movement of tourists and vacationers from Tel-Aviv etc. to the hilly areas of the Lakhish region has started spontaneously, and this movement can be strongly increased by Landscape Planning, for the area is only about 70 km. distant from Tel-Aviv, Jerusalem and Beer-Sheva. In a way this would be a new form for the integration of the region into the life of the country. The stimulation of such development will diversify the sources of income. The Lakhish region is rich in archeological and historical sites, but the promotion of these "resources" depends on making them accessible in a landscape in which circulation and sojourn are agreable even in the hot season.

In a planned landscape development the fulfillment of social and environmental desiderata must be coordinated with ecological considerations. As far as possible, the means applied should serve both social and biological improvement. Most of the areas of afforestation and plantation, if properly planned, will also be important as wind breaks and soil conservation measures; they will form the habitat of many species of birds and animals which feed on agricultural pests, and improve micro-climatic conditions. The consideration of a whole range of social, economic, biological and environmental factors will determine the final decisions on the choice of plantations areas and of the species of plants and plant-associations to be estblished in the Lakhish landscape. Obviously, the choice of ecologically suitable species will also diminish the problem of landscape maintenance.

The creation of environmental values by landscape development is an integral part of the development of natural resources of a region. On the basis of detailed surveys and investigations, the Landscape Plan for the Lakhish region will determine the following uses of land:

Protective plantations around and within the villages and the town. Detailed planning and landscaping of the regional net of communication (for automobiles, carts, bicycles and pedestrians), to connect villages, community centres and the urban centre by attractive alleys. The system of movement and sojourn of tourists and vacationers, including camping grounds, picnic areas, recreational areas and other amenities. The integration of archeological and historic sites in a network of planted strips and pathways. Plantations serving as soil conservation measures. Tree planting as windbreaks on field borders. Afforestation of areas which are unsuitable for food production, including the banks of watercourses and lakes. Nature reserves which interest from the point of view of wild life and vegetation.

REFERENCES

1 See: "To Make a New Countryside" by the author, published in *Landscape*, Spring 1965, Volume 14, No 3.
2 See: *Crete Development Plan 1965 - 1975* (Draft) by Agridev Ltd., April 1965, Volume I, page 177-199: "Sketchplan for the regional development of Crete," by the author.
3 See: *Two Case Studies of Rural Planning and Development in Israel,* by the author, published by Ministry of Housing, State of Israel, May 1964.

VII

TOWARDS REGIONAL LANDSCAPE DESIGN

This article is written in 1954 in honor of Benton MacKaye, on the occasion of the 75th birth of this American forester, conservator, geographer and thinker Mac Kaye is the originator of the Appalachian Trail, a pioneer of Regional Planning, the author of "The New Exploration: A Philosophy of Regional Planning" (1928), and, not least, the outstanding interpreter in his generation of the applied science of Geotechnics, which in his words "...is our emulation of nature in her successful effort to make the earth more habitable."

In a logical sequence, architects have proceeded from the design of individual houses for the family or the small community to planning the environment of these basic cells: street, square, park, the neighborhood, the town as a whole. From here a further step has led to the problems of whole regions. It gives hope for the future of architecture, that architects overcame the boundaries of their professional specialism, collaborated with sociologists, economists and geographers, and began to study the larger geographical environment in order to find the right form and location of the smaller architectural object of the people's homes and amenities. Today a concern for Planning has become in fact a condition for the development of a new architecture of social concern and quality.

In this exploration of contemporary quarters, cities, and regions one discovers that in these larger spatial units relations between form and function, environment and contents of life, have become as undefined, unbalanced, and even untenable, as in the basic unit of the family house. Thus Planning has developed with its important specific achievement a new conception of the town, its life as a social and economic entity, its form, culture, and coherence. Ideas of new values for our towns, of a new urban "art of living" have been created and applied to first realizations.

However, regional planners are detailed still in the preliminary stage of learning from the regional survey and analysis how to approach the specificalty urban, part of the environment. Regional design is still a nebulous notion for laymen and professionalists. Regional Planning in countries such as The Netherlands, Great Britain, U.S.A., India, Israel etc. has found a certain measure of application in the location of towns and villages, institutions, roads etc. Regional landscape design has been left mostly to agronomic technicians and hydro-electric engineers. Even in the probably most comprehensive regional enterprise of our time, the Dutch North-East Polder, regional planning did not rise to the level of shaping a new and better countryside. It is, however, just the appearance of the contemporary cultural landscape which rouses our deep dissatisfaction and even suspicion in respect to the doings of our civilization.

Some characteristics of our cultural landscape are: its rigorous division into areas of full economic exploitation, steamlined to fit mechanized food and timber-production, on the one hand; romantic remainders of ancient fields, hedges, groups of trees and old farm houses, attractive for the passer-by, but too often indicating rural decay, on the other; small pieces of land acquired by townspeople and governments and reserved for holiday makers; and in many countries also areas devasted by exploitive agricultural methods and subsequently abandoned, have become a feature of the landscape of our time.

Our reaction to this deteriorating situation is often a dualistic mixture of resignation to so-called practical needs and romanticism. Our time has transformed the rural landscape into what the Dutch call the "steppe of culture"; water courses have been polluted and their embankments have been denuded of vegetation; what Benton MacKaye calls the "metropolitan invasion" has left its mark of destruction nearly everywhere. Lovers of nature have fled from this reality into anxiously preserved reservations, and, where still possible, into nature far away from concentrations of population. Planners and architects belong to the most ardent defenders of such nature reservations. But until recently, we found only weak arguments for demanding a comprehensive change of the "steppe of culture" into a wholly enjoyable environment for daily life.

This attitude reminds one of the position of those townplanners who, accepting the evils of a 19th century city as an unavoidable necessity, tried to make up for it by designing romantic suburbs, islands of better living ,at least for a few fortunate inhabitants during a few hours of leisure. By now we know how limited a success these attempts at curing the diseases of city life have had. A Planner's conscience cannot be appeased by such half-measures.

The underlying reason for our failure hitherto in design on the regional level has to be found in the lack of a comprehensive approach to the problem of reconstructing the regional landscape. It is true, that our work has made genuine progress when we have made use of Geddes' interrelated trinity of Folk, Work and Place. We often try to measure the social and economic needs and interests of urban and rural commities and to express the results in our physical schemes. However when it comes to the needs of the land itself, we mostly make the mistake of acknowledging only the physiographic character of the factor Place. In fact we consider the properties of a region (Place) just as physiographic *opportunities,* i.e. subordinated factors among which we make our choice and which we use and treat according to our various needs. We have neglected to see in the physical regional environment the home of the most intricate biotic relations and processes of nature, in which man is only one partner. Man's existence depends finally on the maintenance of the biotic partnership with other organisms of the regional environment, that is on its wholeness and health. We may ask ourselves, therefore, if we have not to include in our regional studies also the subject of the nature, the interests and needs of a *regional landscape per se* – a landscape not only of physiographic characteristics, but a biotic community too.

At first this will appear problematic: Need we be concerned with any other interests and values than those of human beings? In fact we are reckoning with such "alien" interests when we acknowledge today the demands "inherent in Technical Development *per se*" – attaching to such statements the faint hope that Technics will reward us one day by easy living. Human needs and values have indeed to be related to factors outside humanity; but among the various possibilities of orientation, biotic processes and communities alive in our environment seem to represent a more rational object of such relations than Technics. It is moreover desirable as well as definitely imaginable, that we unite our own social and economic concerns with those of our natural environment and coordinate our own scale of human values with geological values; for the stability of our social and economic life depends on the state of balance of the natural factors in our environment.

This deliberation does not lead us to metaphysics but to science at its best. In the last decades a most important development in biology has brought us to recognize that the problem of the maintenance of human society on the level of culture is closely connected – if not identical – with the problem of the maintenance of healthy biotic communities in nature. Many people will agree for esthetic or recreational reasons on the need for restoring balanced organic life to our surrounding landscape. But there is much more for us in landscape than areas for recreation and esthetic satisfaction. Soil-science, ecology, conservation, organic agriculture have shown that reconstruction

82 TOWARDS REGIONAL LANDSCAPE DESIGN

of landscape is a condition for a permanently sustained, sufficient, and valuable supply of food and water, as well as for climate and habitability of environment, in a majority of countries. The way in which this unity of interests of mankind and biotic communities can be realized, is our most important clue, when we come to work on regional landscape design.

Planners have experienced the need to become "Jacks of all trades," that is, masters of ever more "subjects," in order to rise to the level of the practical tasks before them. The progress in the field of biology is all the more interesting for us, for it has led biologists to follow the same trend of seeking to comprehend ever-extending fields of life and knowledge. Here new and important partners have appeared for our planning teams. We now have certain scientific fixed points, which clarify the conditions – once we make the decision to plan not only for needs of the moment or quick profits, but for future generations too – that is for the permanence of human culture. From here we may advance towards a well-founded demand for comprehensive reconstruction of our cultural landscape, including its mountains, hills, plains and valleys, its rivers, lines of communication, villages and finally also its industrial zones and towns.

Only a short and incomplete outline of comprehensive regional landscape design, as based on the above mentioned conceptions, can be given here. Our first task will be the definition of organic landscape units, and here we may safely turn to Geddes' concept of the river basin, this "microcosm of nature, seat of man and theatre of history." But we shall have to consider the valley not only as a determining factor, indicating "... the legitimate Eutopia possible in the actual city and characteristic of it," but also as the actual subject of our work of reconstruction. Man in fact has changed the landscape of the river basins by exploitation and subdivision to such an extent, that in turn its characteristic determining influences on man, village and city have become different from what they were in the past. With soil erosion in progress, society too is "eroding," economically and socially. Varying the oft-quoted words of Emerson, we may say that at first man interferes with his landscape and then the landscape interferences with man's life. Mountains denuded of their forests, rivers polluted and progressive soil erosion exert a deeper influence on the long-range fate of mankind than many political movements, promoted by social and economic interests. But it is also in our hands to reconstruct the landscape.

The river basin is an ecological unit. It has been formed by the flow of water and it lives on with the subtle process of regeneration of its resources of water and fertile soil. The state of land, of water, flood or drought, micro-

climate or even macro-climate and power supply in the lower parts of the
river valley depend on the maintenance of biological equilibrium, fertility
and water-absorptive capacity of the soils of the upper watershed line areas.
In the words of the Food and Agriculture Organization of the United Nations, ". . . A nation cannot conserve its true cropland soils and ignore its
maintains and forests . . . the lands of any . . . region are an indivisible unit
. . . " Taking E. Pfeiffer's statement, that a "people conscious of its hills is independent" as a demand stemming from science, we arrive at a further interpretation of Geddes' valley section: it is the duty and the task of the whole
population of a river basin to care for its maintenance as a healthy ecological
unit.

The next step after the definition of natural landscape units will be a
land classification, consisting of two different surveys. One is a classification
of actual land use as established by contemporary social and economic influenced and technical skill. The other is a classification of land according
to use capabilities based on its natural characteristics. "This is essential if
the classification is to serve as a basis for the most intensive sustained use
consistent with preservation." (E. H. Graham). The latter method of land
classification, developed in the U.S.A. in the last 10 - 15 years, may be
considered an excellent basis for our work of landscape reconstruction. We
now have scientific data to help plan for coordination and equilibrium between human technical skill in land use and landscape treatment on the one
hand, and the biological capability of landscape as a permanent source of
livelihood on the other. At the same time we have got a new and firm basis
for subdivision of land according to its natural properties.

The land-use capability classification make possible the transition from
the maximum use of land to its optimum use. It defines the areas which have
to be left to wild life; areas to be afforested and the measure of restrictions
on logging; the areas of permanent pasture or occasional grazing; the slopes
where cultivation is permissible only on terraces, fields where contour plow-
ing, strip cropping or certain crop rotation methods have to be applied;
land highly suitable for intensive cultivation, irrigated or dry farming etc.
etc. This work will have to be supplemented by studies of the natural vegetation and animal life, the hydrologic cycles of the region, possibilities of
power generation, a classification of land for irrigation purposes, etc. Hereby
a new quality of land use becomes possible; its scale is the biological health
of the land, meaning also its permanence as a productive source.

In ancient agriculture, land was used largely in adaptation to the natural
characteristics of the soils. Probably such a practice was achieved by the
trial and error method, but its results, insofar as the face of the cultural

landscape is concerned, may have been in some way similar to what we expect from the application of the new scientific land-use classification. Thus our landscape will regain a more "indigenous" character, although a character modified considerably by the development of our tools and society. In the design of a reconstructed landscape, natural, technical and social facts must be digested and balanced. Indeed, the amalgamation of natural principles with socio-economic principles may lead us to new comprehensive ecological principles of land use. Though their application does not mean a return to the landscape of past times, it is a definite turning away from the "steppe of culture" towards a new pattern, designed by man on the basis of scientifically recognized organic functions.

Such a "New Exploration" of our environment will influence the detailed layout of fields, farms, roads as well. In the contemporary rural landscape, rigid forms of plots and roads prevail, they are convenient for the tractor-driver as well as the land surveyor – though they make our environment even more hideous. But a thorough land classification, surveying the soil in detail, indicates that a rectangular pattern of parcels and fields is justified only in quite exceptional cases. Even in a flat country such as The Netherlands, a more liberal pattern of farms would not be artificial but would lead to an adjustment of cultivation to actual soil properties, to a better location of the farmhouses and a more equal distribution of different kinds of soils among the farmers. In a reconstructed landscape selected species of trees and shrubs, planted as field borders, will serve as windbreaks; while conservation measures will also support wild plants and wild life of ecological importance. Experience has shown that the ensuing increase of crops on such enclosed fields will more than make up for the loss of a certain proportion of arable land. Again, green strips will be planted along roads, railways, wadis and riverlets, not only for esthetic reasons, but at the same time for soil conservation, waterabsorption and the enhancement of the biotic energy of the soil communities. The whole landscape will be subdivided into cells of biotic health and human scale, replacing the desert-like character of a landscape deformed by the contemporary approach to land and cultivation.

The foregoing description of some principles of landscape reconstruction is certainly not equally valid for regions of differing physical, biological and social character: the theoretical and empirical realization of the special characteristics of regions belongs at the very beginning of the actual work of regional planning. It is even more difficult to anticipate the possible influence of the conception of regional landscape design on location and layout of village, town and industries. Doubtless our work will remain incomplete, so long as design does not integrate town and landscape. We have made an important first step by recognizing the primary economic value of arable land; but the imperative of defending arable land against urban expansion does not yet lead regional design to positive trends. We have to do more than avoid and prohibit. We have to relate the pattern of a reconstructed landscape to the new pattern of towns situated within that region. In the period of colonization, the oversimplified land surveyors' design for the American prairie states, the quarter-section or checkerboard system of land-division, led directly to the gridiron pattern of towns and villages. By subsequent subdivisions landscape and town design were indeed interrelated – though damaging both rural and urban environment. In the reconstructed region the relation between town and landscape must be achieved on a higher level. Town and village design will be influenced by the knowledge about its biological and physiographic character and the needs of the zones to which they belong: hills or mountains, plains or valleys, deciduous or coniferous forest, terraced country, extensive or intensive cultivation, sea dunes or river systems. The technical and esthetic integration of towns with these types of landscape after reconstruction, will lead to variegated organic solutions.

Many Townplanners in our days have shown a remarkable predilection for curved roads, irregular space and broken perspectives; this trend sometimes even threatens to become formalistic. By adjusting our designs to conditions of soil, vegetation, topography and drainage, as well as to the specific pattern of their regional environment, we may arrive at new functional foundations for our designs. One illustrative point may be mentioned here: by consuming fresh water and removing waste water, towns and villages exert a large influence on the land of their environment. Especially on sloped territory the solution of drainage problems may lead to the fostering of dense natural vegetation on certain contour-lines and in creeks – and lead hereby to such interesting consequences as the subdivision of a town into neighbourhood units. When approached from the point of view of technics, soil science and biotic environment at the same time, the integration of town and Region will bring forth hitherto unused resources. A town, designed in full consideration of the well-being of its citizens, as well as of health and beauty of its regional environment, would become an important step on our way towards creating a new urban pattern, modified by and improved by the larger regional design.

Since the beginning of the new movement of town planning, planners have experienced the enormous difficulties of convincing governments and the public of the urgency of a new approach to town construction and reconstruction. Doubtless such difficulties are even larger in the field of regional planning, for there more interests are affected and more coordinated efforts demanded. But even preliminary surveys of the state of the land and food production, as summarized by the UNO, are enough to show that in a majority of countries Regional Reconstruction, comprehending landscape and settlements, is necessary for the sake of survival. The reconstruction of country and town, conceived as a permanent reliable resource of life and human environment, means the establishment of a new ecological balance: mutuality in the relations between man and nature. Our landscape needs a new quality, biotic and esthetic at the same time. To achieve that environmental quality, which ensures our survival, a change of attitude is needed, which may best be described in the words of the ecologist Aldo Leopold: "We abuse land because we regard it as a commodity belonging to us." A new and better pattern of regions may be designed and realized, once we begin to see *"land as a community to which we belong."*

VIII

NOTES ON REGIONAL PLANNING AND
TECHNOLOGICAL PROGRESS

The Problem of Integration of Technological and Regional Developments

To the historian technology would appear to play a predominantly constructive role in the formulation of cultures. The techniques of irrigated cultivation, biological and hydrological knowledge, terrace building, the hoe and the digging stick, represent for instance, the prerequisites to neolithic culture in many parts of the world. The regional pattern of settlement in medieval times, to cite another example, distances between villages and towns, the division of labour and rural-urban socio-economic relations are closely related to the technological level of the means of communication, road construction and production at that particular period.

Actual experience, however, forces us to realize that technological progress in our time contains disruptive potentialities. As a consequence of the introduction of mechanical power and mechanized tools of production and communication, we witness a major deterioration in the functions and forms of settlement and landscape. Technological development has not led of itself to a new era of socio-environmental harmony. Urban and Regional Planning can be considered as an attempt to assimilate the new technology into a new environmental culture. Planners are unanimously striving for a thoroughgoing reconstruction of the environment of work, rest and leisure, in order to achieve that integration.

Underneath this apparent unanimity, however, decisive questions of Regional Planning policy remain hidden and therefore unresolved:

Should Regional Planning be guided by trends of technological progress and aim at the "streamlining" of the environment in accordance with the functional patterns called for by the differing character of steam, electrical or atomic energy, respectively, or increased surface- and air-travel, or automation?

Or should Regional Planning be guided by broad socio-environmental goals, leading us to select those phases of new and foreseeable technological developments which are useful for the achievement of such goals?

It will be obvious from the outset that the first alternative presents us with very clear and easily conceivable starting points, whereas the second involves Regional Planning in the necessity to consider a great number of factors and problems which vary from locality to locality. But such difficulties do not make the issue any the less real. It appears that a Planning body cannot avoid taking sides in this issue when clarifying the problems and goals of a Regional Development Scheme; in fact, every Regional Scheme contains, implicitly or explicitly, decisions about the role to be assigned to technological advance in Regional Development.

Technological Determinism

Several authors have noted the astonishing fact that the 19th century, in spite of all the revolutionary changes it introduced in to the environment, neverthelss adhered to an environmental determinism, explaining the patterns of settlement and society as a result of the influence of geographical factors and environmental adaptations undertaken in the course of man's fight for survival.

Nature seemed to justify *laissez-faire* methods, and was held ultimately responsible for its socio-environmental consequences. The idea of man as an active geographic factor was relegated to the specialized technician (1A) (2).

This attitude which reflects the blind optimism of that period and its complete neglect of any but technological planning, has been rejected in our time. Man is now recognized as an activating factor of prime importance; but we have been quick to transfer the "determining powers" from climate, water, topography, etc. and to invest them in technological progress as such. Technological progress though created by man, is now thought to be an irresistible process – a quasi-natural evolution, bearing inherent values and to be pursued vigorously by every progressive population.

Therefore it is the "order of the machine" in which the comprehensive designer must now find his place" (3A) (4).

Though only few wholly technologically determined regional plans have been carried out until now, the idea of "techno-determinism" has a considerable influence on the methods of Regional Planning; it is a deeply rooted attitude, wide-spread among technicians, scientists and planners, and cutting across political and social borders. It might be interpreted as part of the ideology of the so-called "Managerial Revolution" (5). Regional Planning, in this conception, becomes "total" planning, and is assigned the task of making environment and society conform to the evolution of mechanical power and tools.

A highly charactristic case in point is linear development, or the linear city. It was proposed during the first Soviet Five Year Plan for Regional and Urban Development in Russia. Similar ideas have been developed in the United States on a gigantic scale. Le Corbusier's linear city, for instance, is characterized by continuous parallel lines of railways and a main highway, on one side of which are threaded industrial units of production, and on the other side stripe-like residential units. This idea offers excellent conditions for industrial transportation, horizontal expansion and combination of production processes. The proposal infers, on the other hand, a completely disintegrated urban structure, by turning, as it does, the residential part into an endless tape divided into small units; each residential unit is directly attached to an industrial plant, as though were a part of its equipment. Any arrangements allowing for urban diversification, the creation of a sizable town center and a possible change of occupation which does not involve a change of residence, have been completely neglected.

The enthusiasm for the new technics of communication provides Le Corbusier with another starting point for Regional Planning. Route Planning is in his view the backbone of the new Regional Plan. By harmoniously relating the activities of towns and regions with the four routes – the earth, the sea, the iron and the air route – he envisages the planner deciding the destiny and changing the pattern of settlement of the technical civilization.

"How cheerfully we shall travel the four routes towards a brave new world" (6) (7).

The lay-out of the so-called "Dispersion Cities" in Russia was even more drastically determined by industrialization; they were "grouped around industrial combines and surrounded by agricultural zones and truck farms," with a view of identifying urban centers "with the creation of a specific industrial enterprise, the nature of which would be indicative of the character of the resulting urban pattern." Indeed in a decree issued in 1933, the tasks of Regional Planning were estblished as the providing for the location of facilities in the following order of priorities: Industry, Power Stations, Specialized Agricultural Areas, Transportation, Administration, Water-Supply, Sewage Disposal and Forest Areas – with the location of settlements not even mentioned. Regional Planning in the East-European countries, on linear or other principles, had an additional aim: it was "moving towards the liquidation of the differences between the city and the villages," with the effect of a widespread urbanization of the cultural landscape. The result of these planning directives was an urban dreariness not very different from that encountered in the unplanned West-European industrial towns of the 19th century. (8). (9). (1B).

A different type of a technologically determined planning is now being advocated in American and European metropolitan centers, where growing land-needs and traffic congestion have reached a critical stage.

The movement and containment of traffic, with the main emphasis on the private car, has been accepted as the key problem of Town and Regional Planning; as a recent issue of "Forum" declared, car-density has now become "the crucial figure for United States Planning." (10).

The proposals made on this score would in effect dissolve the central urban structure, by stretching it along a highway, by "islanding" the center, which would consist of a cluster of skyscrapers in the middle of a sea of parking areas and car approaches –or by assisting its explosion by the dispersal of central institutions to the countryside, where they would become accessible only to long-distance travel.

But the sub-urbanization of the country and the emigration of central urban facilities are advancing even without such plans, and their realization would only demonstrate in the most expensive way that the metropolitan city has transgressed the limit of possible concentration of functions and populations. It seems, therefore, that in the long run technologically determined planning cannot replace the search for comprehensive, constructive solutions on a regional scale.

Technological Progress towards Mobility

The assumption that technological progress contains self-acting, inherent trends determining the fate of mankind, should be rejected. Even the horrors of "paleotechnics" are not a result of the quality of its technological standards. Whatever the outcome of technological advance, it is man who realizes his own ideologies and intentions through the medium of technology. It is nevertheless true that such ideologies are not always explicit, or even clearly conceived. The fact that the integration of the machine in regional structure is causing such great difficulties is due to profound reasons. With Regional Planning we tend to attain a form of equilibrium between society and environment, which enables man to settle down in optimal locations and organize in well-balanced settlement structures. Modern Technologey, however, has meant for us, in general, the use of science and natural resources to achieve an ever growing mobility of man – a sort of a new nomadic civilization. Progress is conceived as a cumulative process of achieving an ever-greater expansion, skill and the liberation of the individuum from environmental belonging – conditions of space, time and society; "Neotechnics" of the "Second Machine Age" are "better," because they seem to make this escape ever more complete.

The greatest achievements of modern technology, in function and form are not in static structures, but in vehicles, vessels and planes. The linear regional scheme would make their movement more convenient, and it happens to be a direct continuation of the distribution of nomad's settlements in strips along travelling routes – settlements that are half-town half-village (11). Even modern architecture shows a tendency to design tent-like, easily removable and transportable buildings. Automation is freeing industry from several localizing conditions (12) and through the development of atomic energy, power-stations are becoming a footloose industry, which can be set up anywhere, since they can exist virtually without fuel shipments. The prospect of a completely transportable industrial development is rising into view. One might claim that from the point of view of man as a settler of the land, the objectives of the most representative achievements of this technology often appear irrational. But man as world traveller is offered superior equipment indeed.

The Ideology of Regionalism

Life, as most exponents of Regional Planning conceive it, is characterized by being "essentially social" (13). Every living creature, including man, exists by occupying a functional position within a structure of inter-related living beings. (1C). Health and Wholeness, the goal of human life, can only be attained by the continuous regeneration of body and mind, nurtured on fertile and self-renewing resources and environment, as well as on harmonic cycles of give-and-take in the relations between men. To enhance its vital forces, humanity according to this conception, should strive towards more integrated equilibrium-systems. To tackle the problems of population increase and the long-term use of natural resources, Regional Planning envisages new settlement-structures, that provide for a re-intergration of man and environment for a long time to come.

In Regional Planning, life-values are viewed as accruing from a settled man-environment relationship – a quality which cannot be measured in terms of technological progress. Not maximum singular achievements but optimal mutual relations are aimed at. Such factors as can enhance this equilibrium – stable food and water supplies, as well as other natural resources, general environmental conditions, relationships between neigbouring localities and regions – are looked upon as the most desirable anchorage for human settlement. They are the basis for the growth of the "good life." Traditional regional structures and communities with a clearly defined technology are studied with great interest by the Regional Planner – not because of their primitiveness, but because of the observable wholeness and unity of functions and forms they have attained.

But the admiration for the past is liable to mislead Regional Planners to reject all the achievements of our technology. Their main motive would appear to be the fear that technology will necessarily make regional communities ever more independent of the environmental anchorage, thereby shaking the foundations of all social values. For the same reason, these planners are prone to welcome too uncritically the warnings of the "prophets of doom" against the consequences of an exploitative civilization: the depletion of food and natural resources, unemployment caused by automation, uncontrollable nuclear radiation etc. Technological progress, in this view, finally effects its own destruction. The attainment or preservation of regional settlement structure would therefore necessitate the discouragement of technical innovation. But in point of fact this approach implies the acknowledgement of the irresistible force of technological development, once it has been initiated, and we can therefore assume that this "romantic" form of regionalism is just another form of "techno-determinism."

Technology is Neutral

Any scheme for a regional structure determined in the main by the new technology, seems wrong because the very quality of our technology lies in the fact that it so immensely widens the range of possibilities of human settlement, production and community interaction. Technology having been used for creating extreme centralization of power, population and commodities, and for keeping such settlements alive (though at a high cost) could also serve the decentralization of population on a high level of efficiency and convenience. New sources of power make it possible to supply highly complex services even in small rural settlements. The town of Emmeloord, the urban centre of the newly reclaimed Dutch North-East Polder, proves that it is now possible to maintain a high standard of services in a small town of less than 10,000 inhabitants. While the concentration of industries may cause the social and physical evils of congestion, dreariness and economic lop-sidedness, electric power increase the chances of decentralized industry, and thereby strengthens the socio-economic structure of agricultural regions.

Tractors and combines may be used to hasten ruthless soil exploitation and erosion. But similar machines have tremendously increased our capacity to reclaim and conserve the soil; and small mechanized agricultural tools can alleviate drastically the harshness of agricultural work, and can be employed to intensify land-use. Medicine can on the one hand serve to increase the birth-surplus of a population, but on the other hand it may serve to stabilize the population by birth-control. Technology can be used for soil destruction and the disruption of subsistence agriculture, and in this respect it may be considered a *"cul de sac"* phenomenon, similar in effect to the avalanche of equestrian nomadisation in the 12th century (1D); but Technology can also serve the "horizontal expansion" of agricultural settlement, even of subsistence agriculture (3B).

It is not incumbent on modern development to be stretched or "clustered" along highways; trains, cars or helicopters should rather serve to make accessible even the remotest spot where environment or resources attract human settlement. The question is frequently discussed whether the development of communications make communities more interdependent or more independent. As a matter of fact both possibilities have become increasingly realizable by technological developments. The choice between them has therefore to be made from other than technological points of view; it has to follow the plans of social organization and environmental reconstruction. Wherever technology is introduced or developed, the populace or the Development Authorities will have to decide between such alternatives.

The Regional Use of Technology

Technology which has so often been applied in an uncoordinated manner, being considered either as the motive force or as an obstacle to the success of regional developments, must be integrated within a comprehensive Planning Concept. The destiny of Regional Planning is to help the developing regions to avoid the negative social and physical consequences of industrialization, to replace the impact of unplanned technological development by a comprehensively planned regional evolution. The global expansion of a typically Western technology can be explained by its political, demographic and socio-economic background; the use of colonial resources and markets; European population pressure and Capitalism.

But the developing regions of Asia and Africa have proved, at a heavy cost, the need for a critical approach to the use of Western technology.

On the other hand it is clear that the realization of large scale Regional Development Schemes in our times can not be made feasible without the use of modern technology. Planning and development for the benefit of man mean, therefore, replacing the universal mechanical application of formulas for technological progress, developed according to European and American needs, by "Regional Technics." (15). Such regional technics would be adjusted to the climate resources, landscape, society and culture of the region concerned, and would be based on a man-environment relationship, which by definition is different in each region and forms a different starting point for Regional Development.

No universal "normal" attitude towards natural resources, conceptions of needs, degrees of specialization, urbanization etc. actually exists. (16). Labor saving machinery which is suitable for certain sectors of the United States economy constitutes a very problematic factor when applied in an overpopulated country without full consideration of its social, physical and economic implications. Adequate labor-using, in such cases, may be more important than labor-saving. Thorough surveys are needed to reveal the essential problems of the man-environment relationship prevailing in a given region; and the socio-economic and physical goals of development have to be defined realistically, so that the most appropriate technological devices can be selected and employed.

The case of India is the most interesting example of a new attitude towards the role of technology in Regional Development. Gandhi did not aim at a rejection of industry, but at its selective introduction into Indian society, without upsetting its pattern of agriculture, its principles of social justice and the cultural values of "Truth, Non-Violence, Total Awakening, Dignity of Labour and Non-Possession." For this purpose, "machines would be of special type suitable for removing drudgery of handicraft and improving its quality."

Big industries would be owned by the State, to prevent their use for purposes of individual greed and profit, and all would operate in the general context of a decentralizing industry (17) (18). India today is developing largescale modern factories, as well as its wide-spread traditional cottage industries, with plans for future electrically-driven looms in the cottages of the weavers (19). This is the result of planning the "major social economies" of industrial and urban development, which include the economic and social cost. (3C).

The social and geographical integration of the industrialization process is here conceived as a function of the regional approach to Planning. (18). The Indian Community Development Projects, already covering 25,000 villages with a population of 16 millions, are creating by socio-economic and physical organization and education a regional countryside structure which aims to develop agriculture, markets, health, education, services and industries in a comprehensive and harmonious manner.

The plain Indian and Pakistan statements, to the effect that the purpose of technological progress is to provide "more satisfying lives" (20) to the people, are important and hopeful indications. But it would be a mistake to assume that such attitudes are characteristic only of Far-Eastern culture (14). The first Western planned regional developments have been conceived in a similar spirit, aiming at the realization of social ideals, technology being put at the service of Regional Settlement. David Lilienthal considered the benefits of the T.V.A. not merely in terms of economic and technological expansion, but in the fulfilment of the desire of men to have roots in some place and to build "spiritual strength" for themselves. (21). The permanent balanced use of land is the goal of the organization of soil conservation districts in the United States, covering 85% of the agricultural area of the country and employing scientific methods and a specially developed technology. (22).

In Holland the highest achievements of hydrology and other techniques are serving the reclamation and development of new land, strengthening and extending the rural regional structure. In Israel the planning of modern agricultural settlements and the integration or rural-urban development was undertaken with a view to the intrinsic values of a settled man-environment relationship, especially for a rootless and mostly urban population. The social scale of these and similar enterprises is even smaller than the beginnings of the Community Development Project in India, but they should be considered as pilot projects making possible the "horizontal expansion" as well as variations of Regional Development and its specific technics.

The Progress of International Technology

On the international plane, there is need for the adaptation of the general progress of technology and machine production to the needs of balanced human settlement. Once it is acknowledged that the problems of the technologically backwards nations are not exclusively their own and that the advanced countries cannot ignore such backwardness (3D), the direction that the industrial countries give to technological progress becomes a heavy responsibility. In principle there is nothing wrong with the development of automatic and repetitive processes of production of standardized commodities which are equally needed in different parts of the world. But it is a tragic fact that the best efforts of technicians and scientists are still directed at the development of faster planes, helicopter communication, space travel, and various luxury gadgets serving either war or the interests of a minority. It is equally deplorable that a number of Regional Planners have found their vocation in the study of the future metropolitan centers as affected by such peacetime or wartime developments. Technology to date has been employed only to a minor extent in the alleviation of the greatest and most dangerous hardship afflicting the majority of mankind – the lack of food and shelter. (23). The use of technology to reclaim and conserve agricultural land and to intensify its cultivation for the production of healthy and plentiful food, the construction of farms and decent housing and the alleviation of physical toil in production and service, is today in no proportion to genuine needs. The relationship between the industrial countries of the West and the technologically backwards countries still bears the character of a superbly equipped nomadic civilization invading agricultural countries where a subsistence economy and a rural culture obtain. This may partly explain the distrustful attitude of the East towards Western methods and technology. The development of Regional Planning and Soil Conservation in the United States, Holland, and England may be interpreted as the sign of a renewed appreciation of the values of a settled regional structure. The more such developments gain ground in the West, the more the West's comprehension of the Eastern attitude to development problems will grow, and the greater the chance for mutual understanding and trust, for more realistic progress in the use of international technology.

Summary

The introduction of coal, oil, hydro and atomic energy have each been interpreted as initiating and determining a new and better pattern of Regional Development. But it is illusory to assume that such factors of themselves can lead to the improvement of living conditions in a region. Realistic Regional Planning must find its cues in the more comprehensive conceptions of social values, the quality of the man-environment relationship and a functionally and aesthetically satisfactory cultural landscape. Technology has to serve the achievement of desirable degrees of concentration, dispersal and interaction of communities, of the rhythm of the alternation of gregarious and secluded living, and the creation of a balance between the natural and the artificial ingredients of human environment. With such conceptions overshadowed by the prognosis of technological progress, Regional Planning becomes a meaningless and hopeless affair, a lop-sided adaptation of society and environment to conform to mechanization. On the other hand, technology, when integrated in the Regional Planning Concept, may become one of the means of liberating human energies, which were formerly confined to the bare effort of survival, for the enhancement of the quality of life everywhere.

What are the prospects for a regional integration of technological progress? The world-wide and still growing impact of the problems of hunger, population increase, soil erosion, housing, settlement and lack of minimal conveniences in non-European and European countries, may lead to a more realistic "horizontal" expansion of technological progress, to be employed as a means of basic improvement of regional conditions where it is most needed.

Sociological and biological considerations may for the same reason lead to a reappraisal of the values of regional patterns of human settlement which would supersede the values of human mobility.

The main hope for the regional integration of technological progress, however, appears to lie in the spread of a changed attitude: the conception of environment as a bio-social community in evolution, the sympathetic orientation toward the future of mankind and life in general, – and the belief in man as the active agent of change in space and time.

REFERENCES

1. WILLIAM L. THOMAS, Editor, *Man's Role In Changing The Face of The Earth*, An International Symposium, Chicago 1956.
   A. CLARENCE J. GLACKEN, *Changing Ideas of the Habitable World*.
   B. ALBERT E. BURKE, *Influence of Man upon Nature – The Russian View*.
   C. PAUL B. SEARS, *The Processes of Environmental Change by Man*.
   D. H. VON WISSMANN, *On the Role of Nature and Man in Changing the Face of the Dry Belt of Asia*.
2. K. E. BARLOW, *The Discipline of Peace*, London 1942.
3. BERT F. HOSELITZ, Editor, *The Progress of Underdeveloped Areas*, Chicago 1952.
   A. WALTER R. GOLDSCHMIDT, *The International Relations between Cultural Factors and the Acquisition of New Technical Skills*.
   B. KONRAD BEKKER, *The Point IV Program of the United States*.
   C. BERT F. HOSELITZ, Preface.
   D. ALEXANDER GERSCHENKORN, *Economic Backwardness in Historical Perspective*.
4. KONRAD WACHSMANN, "Building in our Time," 1956, (Unpublished Manuscrips).
5. JAMES BURNHAM, *The Managerial Revolution*, London 1945.
6. LE CORBUSIER, *Concerning Town Planning*, London, 1947.
7. LE CORBUSIER, *The Four Routes*, London 1947.
8. MAURICE P. PARKINS, *City Planning in Soviet Russia*, Chicago 1953.
9. ACT AND GOVERNMENT MEMORANDUM, The First Czechoslovak Five-Year Plan, Prague 1949.
10. *Architectural Forum*, September 1956, Vol. 105, No. 3, New York.
11. ERNST EGLI, *Climate and Town Districts*, Zurich 1951.

12.   DAVID G. OSBORN, "Automation of Industry – A Geographical Consideration." *Journal of the American Institute of Planners,* Fall 1953.
13.   JORIAN JENKS, *From the Ground up,* London 1950.
14.   See the works of Emerson, Geddes, Mumford, G. T. Wrench etc. etc.
15.   KARL KRUGER, *Ingenieure Bauen Die Welt,* Berlin 1955.
16.   ECONOMIC COMMISSION FOR ASIA AND THE FAR EAST, Draft Report of the
Seminar on Urbanization in the ECAFE Region, 1956.
17.   *International Social Science Bulletin,* Vol VI, No. 3, 1954, D. P. Mukerji,
"Mahatma Gandhi's Views on Machines and Technology."
18.   CATHERINE BAUER, *Economic Development and Urban Living Conditions,*
Draft 1956.
19.   JOHN STRACHEY, Interview with Nehru, *The New Statesman and Nation,*
June 23, 1956.
20.   UNESCO, *The Social Implications of Industrialization and Urbanization,*
Calcutta 1956, A. F. A. Husain, *Human and Social Impact of Technological Change in East Pakistan.*
21.   DAVID LILIENTHAL, *Democracy on the March.*
22.   U.S. DEPARTMENT OF AGRICULTURE, *Technical Skill for Soil and Water
Conservation,* Foreward by H. H. Bennett, 1950.
23.   UNITED NATIONS, Dept. of Social Affairs, *Preliminary Report on the
World Social Situation,* 1952.

IX

CREATING NEW LAND: DESIGNING ON NEW LAND

It is all too easy to criticize a new landscape and to say that it is good or bad, beautiful or monotonous, and I think different people will react differently when seeing for the first time the landscape of the North-East Polder in the former Zuider Zee. Some of them will say that it is strangely impressive, and others will say that it is unhuman. But I believe that the reason for this vagueness lies in the lack of any scale which would help us to make such evaluations of a regional landscape from the functional and aesthetic point of view. Good and bad must be related to something, and I will try to contribute to the creation of such a scale.

In the development of any old landscape a number of special and temporary conditions, besides the temporary needs of man, contribute decisively in determinig the fate and the shape of the land. The natural, topographical and historical land conditions counterbalance the impact of man's changing ideologies and techniques on the land. They preserve a measure of continuity in the changing landscape, they represent a firm background against which development is seen to be taking place and becoming manifest. For the landscape planner such factors represent an anchorage which safeguards the preservation or the creation of landscape character. Now on new land, reclaimed from the sea, from swamps, from flooded areas or deserts, such a welcome mitigation of man's impact on the landscape is either missing or reduced to a minor role. Here land is a tabula rasa and man's role as a landscape improver is put to a most critical test. The community has to demonstrate in a vacuum its true capability of creating a permanent and workable, habitable and life-favouring environment.

In several countries the reclamation of new land is promoted enthusiastically, but when they come to evaluate the results observers are somewhat dubious about the new character of the landscape created. We have, in fact, been warned beforehand by the designers of the new land that the results of planning a completely artificial modern landscape might be aesthetically

unsatisfactory and that some uniformity and even rigidity in the new landscape pattern could not be avoided if planning was rationally applied. May I
quote a sentence from a very good article, written by Prof. J. T. P. Bijhouwer
about the landscape of the Nort-East Polder, in which he says: "Unfortunately it is generally felt that there will be a shortness of rational means to prevent uniformity and consequently monotony in the polder landscape ..."
Now should we consider this as an inherent limitation of contemporary
landscape development and submit to it? Or should we regard it as warning
that the present planning programs and procedures are faulty?

Let us try to limit the field of discussion on this subject. In dealing with this
question, let us realize that the cultural landscape, as distinct from parks and
pleasure gardens, has never been designed with the primary aim of providing
aesthetic satisfaction. A satisfactory landscape character was a by-product
of the successful solution of spatial and temporal problems of land-use and
management. Yet, on the other hand, it seems that without this by-product
man does not adapt well to any environment. Although it is difficult to prove
it conclusively, we may say that the visual satisfaction that the landscape im-
parts to its inhabitants is, in fact, as vital for a community as the satisfaction
of biological and social needs. Furthermore, it may be stated as a designer's
belief, that biological, social and aesthetic needs form an indivisible whole
and must be satisfied as such.

The cultural landscape, then, cannot be beautified by a separate act of
planning. If modern land-use planning must inevitably lead to rigidity, then
it would seem that we must resign ourselves to the loss of environmental values, hoping that time will perhaps soften the harshness of the marks carved
by us on the land. But then the truth could not and should not be hidden
under pseudo-romantic or artificial nature elements. Nor would an emphasis
on streamlined forms and vastly increased modern scales enhance the attractiveness of a rigid landscape pattern. In landscape design on a regional scale,
even more than in architecture, the creative form must make sense, and
honesty of expression is simultaneously a moral requirement and a pre-condition for durable results. An admission of environmental poverty, therefore,
would be preferable to designing a fake landscape.

Observing the changes of the rural landscape in most countries, the conclusion may be drawn that modern man is not only the conqueror but also
the originator of visual emptiness. The most representative phenomenon of
contemporary landscape development is not the reclamation of deserts or
sea bottoms but the creation of vacant landscapes where there formerly
existed a wholesome human environment. During the past hundred to one
hundred and fifty years our expanding technological civilization has treated

much of the old land as if it were a vacuum. By systematic burning and clear-ing of vegetation, by levelling and rectangular subdivision, the *tabula rasa* for man's exhaustive economy was prepared. By disregarding the ecological chain of biological, topographical, climatic, hydrological and social condi-tions of life in the landscape, land management led finally to the creation of what the Dutch call, so significantly, "the steppe of culture." This widespread and monotonous type of landscape is characterized not only by being clima-tically and aesthetically unattractive but also by functional defects such as water and wind erosion, and unbalanced water cycle and declining soil fer-tility. This landscape constitutes, today, a waste of a much greater extent than the potentially reclaimable water-covered and desert areas.

The deterioration of the old rural landscape and the formation of an un-satisfactory environment on the new land cannot be explained just by what is called our "alienation from nature," or the artificiality of the environment created by us. Let me state my interpretation of the notion "natural." I be-lieve that the concept of the (natural) condition of man does not express a past condition, left behind with the process of civilization, but a balanced relationship between human capabilities and natural forces which can be achieved, renewed or re-established at any level of civilization, that there exist a sufficient amount of wisdom, good will, and energy.

The man-modified landscape is not identical with the unnatural land-scape. I believe the cultural landscape is by definition always a man-modified environment. And yet, in the past, the artificial increase of its functional values for man did not reduce its biological and aesthetic values. There exists, perhaps, the quality of the purpose set by society to land development. In the case of the "steppe of culture" the land use aim is narrowly conceived as a rule, and of temporary nature. It is limited to the maximum exploitation of soil fertility within a minimum of time, and this infers a readiness to aban-don the land after the exhaustion of its reserves as if it were a mine. This attitude results, finally, in the monotonous, desert-like appearance of such a landscape. But the aesthetic deficiencies often encountered in newly re-claimed and settled land may also indicate that the land-use aim was not broadly enough conceived. We can expect the creation of a satisfactory cul-tural landscape when men develop a comprehensive interest in land use. The fact that in the past communities realised that the land had to serve for gene-rations as a cultivable and renewable source of life, as a permanent place of habitation, work, celebration, movement and rest, is the main cause for the sensation of environmental wholeness and aesthetic satisfaction which we ex-perience in a preserved, traditional landscape.

As planners for the future of the land, however, our cardinal aim should not be to preserve or to emulate such landscapes, but to regain the scope which originated them. It would then seem that we still have to explore and to establish what really are the broader purposes of modern landscape development. The most obvious aim for the development of new land lies today in the efficient and intensive production of food and raw materials for the increasing needs of populations. This is a quantitative aim and it should be brought into harmony with the broader spatial and temporal aims of development which we can generally characterize as qualitative. It very often seems, though, that the quantitative aims emerge from immediate, short-range considerations, and that the qualitative aims are related to the future. For the very reason that the development of new land into a cultural land-scape demands great economic sacrifices, long-range interests should be taken into consideration as not less important than the immediate interests. Besides serving the immediate needs for food production, therefore, the three following principles should be incorporate in the planning program of the contemporary landscape.

The first principle: the cultural landscape should be planned for an optimal sustained level of soil fertility. Consideration of the needs of intensive production on the one hand, and of soil protection, vitalization of microbiological soil life and preservation of the water cycle, on the other hand, have all to be taken into account and to be balanced with each other. Once this is accepted, landscape planners should make use of ecology as an applied science. A balanced relationship between man and the land means the achievement of a new ecological climate in the landscape, which is arrived at by a conscious and rational effort of man. Land development and management should be based on the study of the structure and history of the local landscape, on meticulous soil capability classification, on the top-soil map, the geological map, and so on. Ecology, I repeat, thus becomes an applied science, and sustained soil productivity as a cardinal purpose of cultivation should become a major landscape shaping factor overriding considerations of a more temporary character.

The second point is the need for a balanced social life. In planning the contemporary cultural landscape, it should be taken into account that the standard of living of a modern rural population in industrialized countries must be adjusted to that of the urban population, though rural life will always represent a different kind of life. It is, therefore, another function of the cultural landscape to provide attractive environmental, social and cultural amenities to its permanent inhabitants. In this respect the habitability of a landscape is also a means of maintaining its productivity. Planning for contemporary rural habitability is still in a preliminary stage. New factors are constantly emerging, influencing the distribution, location, size and degree of concentration of villages and village services, determining, in short, the whole settlement structure of a regional landscape. This structure, as well as the road network, power-lines, and other services must be related to contemporary means of communication and methods of production, distribution, and vehicular traffic. In some parts of Israel, such as the Lachish region, the beginnings of such a new, adjusted pattern of habitable landscape can already be discerned, though, I repeat, only the beginnings.

The third point to be observed is that the general increase of leisure and of demands for more first-hand experience of different environments, combined with the availability of faster and cheaper means of communication for the masses have revolutionized the mobility of the whole population of industrialized regions. This new factor is liable to cause fundamental changes in our conception of the rural landscape. Today this free, so to speak, uneconomic movement is still in its very beginnings, but it is rapidly gathering momentum. A working week of three and a half days, which may become the rule in some countries, in the near future, will turn the population into potential modern semi-nomads. Their wandering cannot and should not be restricted to the highways and the advertised, overcrowded resorts. The movement will be largely directed towards areas of functional interest and it may greatly increase the direct contact between the urban population and the cultural landscape and its inhabitants. I believe that if there is any room at all for an optimistic view towards the future, one may forecast a rise of the urge for direct biological and social contacts resulting, among other things, in a return movement of urbanites to the landscape. We should not regard this development as a new and overwhelming danger for the landscape; it would have little in common with Rousseau's desperate return to nature, for it would involve the rediscovery, reconstruction, recreation of the cultural landscape with the active participation of the urban population. We should meet this development by planning an environment both accessible and hospitable to what may be called the recreational movement, recreational in both the passive and the active sense of the word.

The effect of the introduction of such aims as the three just mentioned into landscape design can be observed in the new landscape of the Tennessee Valley where a "steppe of culture" has been transformed into a fertile, habitable and hospitable environment. It can be foreseen that such a diversified landscape development program will entail an honest differentiation of the landscape pattern on new land too. Mere economic, technological land planning is an impracticable and obsolete means of achieving such long range ends of environmental development. It also means that the division of a region into separate land-use zones, for food production, settlement and recreation separately, is a most undesirable solution of the problem. The most suitable and enjoyable resorts are not the specialized recreational areas but the farming and fishing villages, the places of living and working. The urban and rural population should not, and in most cases cannot, be herded into separate enclosures. Multi-purpose land use programs, therefore, may offer the best solution for the emerging environmental problems of our time. They can be considered as an attempt to return to the comprehensive conception of land-use aims which serve as a basis for the creation of a liveable and enjoyable contemporary environment.

I would like to mention some possible, practical measures for realizing these general purposes. Most of these are, of course, not new inventions but they are practices which are applied in Holland and in many other parts of the world. Such practical measures for the design and planning of new land would include the following. First, that group of measures which perhaps belongs mainly to the first point which I made: to the creation of sustained fertility in the landscape. The layout of plots, fields and forests, the crops to be cultivated, the size of the farms, the tracing of roads, the borders of fields, the location and shape of built-up areas – in short, all the components of land use – must be determined by taking into full consideration scientifically defined differences, even subtle differences, in soil quality and topography.

Second consideration would be that wooded or otherwise planted green areas and water covered areas should be developed to create an ecological balance in the landscape, especially to keep down pests, but also as wind breaks and as anti-erosion measures. Another type of such measures would be aimed at enhancing the habitability of the cultural landscape: forests and green areas should be planted to improve the physical and the social climate and health conditions, and to create protected open spaces. The types of farms and cultivation should vary according to their distances from the population centres.

A third group would be planning for scientific and cultural development; of such measures I wish to mention two. The first is that uncultivated and untreated stretches of land should be preserved as laboratories for ecological research. And the other is that specific kinds of flora and fauna, which inhabited the reclaimed area in the past, should be protected in parks or in other chosen spots for educational, cultural, and scientific purposes.

Now, coming to the important point of the increased movement of whole populations in our time, what would be typical, practical measures to meet this development? Certain areas in a new landscape should be reserved for the sojourn and rest of a temporary, visiting population. Roads should be designed and roadsides treated with a view to the fact that the roadside represents the most critical line of contact between the traveller and the landscape. The landscape should accompany the real movements of individuals and communities in a positive manner. In general the landscape problems should be subjected to rational analysis serving as basis for a creative design which aims at the lucidity and wholeness of the layout, with no attempt made to hide behind traditional patterns or to camouflage the artificiality of a landscape developed in a vacuum.

In discussing landscape design in general, it seems more appropriate to enlarge on the programatic basis of design than to deal with the principles of design proper. Not all the stages of the processes of design lend themselves equally well to rational conception and useful discussion. Besides, landscape design must always be seen as related to a specific set of conditions — human or other conditions — which decisively influence the designer's conception. Rules for landscape design cannot therefore be rigid and the discussion of the quality of design must be related to specific cases. We can postulate, however, in which climate the new art of landscape design on a regional scale would grow best; and to this end we must define useful concepts of the procedure of such design and of its progress. In environmental design we create a mold for dynamic life processes, and we might perhaps do well to conceive of it as an emulation of morphological processes in nature. We observe, in nature, that a tree, a plant, an insect or an animal is adapted to fulfil a great variety of different functions: even each of the individual parts of an organism, maintains its own particular relationship with the external environment. Yet the organism is characterized as a single individualized whole because it possesses its own specific, innate architectonic. Similarly we see that the hundreds of different traits of a traditional town or a region have grown by a process of trial and error into a balanced structure, unified in history by a form singular to it. It is such a unity of form, character, or architectonic that we are trying to create in the contemporary cultural landscape by integrated design. Rural and urban developments of the recent past have shown that the simplification of the Land-use programs and the consequent shaping of the environment by the mere multiplication of simplified forms, such as a house or a farm or a field, will not reach that level. It will lead to an amorphous uniformity in the landscape, but not to formal unity. The essence of the latter is both variety in unity and unity in variety.

Observation of the natural and historic morphological processes may lead us to distinguish three stages in the process of landscape design. The first perhaps, is the statement of the basic purposes of the plan and, consequently, the broadening of the program by taking into account all these variegated demands of sustained fertility, habitability, accessibility, and hospitality to which I have already referred. The second dimension – if you wish to say this of the process of planning – would be the coordination and synthesis of the various, sometimes even conflicting, functional land-use needs in order to make comprehensive development feasible at all. And finally, the third part or dimension would be the transformation of the rational synthesis into design of a distinct quality and character.

You would thus have a steady progress from scientific analytical thinking to rational synthesis and to creative design. In such a process the design work would be carried on alternately by teams of specialists and by individual designers. The designers would be aware of the broad biosocial basis of their work, while the teams and administrators would recognize that only individual designers can be trusted to realize unity in variety. Once we employ a broader planning approach we might feel more confident of the landscape forms, geometrical, regular or other, which we are creating. It seems certain that we shall not succeed in further humanizing – if we may use this word – the anyhow man-made landscape on new land by introducing plants or architectural forms of the natural, rural history on the old land. A design based on a full realization of the environmental needs of our communities holds the promise of better results. This has been repeated in our day, time and again, but we believe that once our spatial and temporal aims in the new landscape are broadly enough conceived we shall feel less inhibited when designing the landscape on a regional scale, and thus we shall create in a straightforward manner the architecture of the new land.

Now in no country can the progressive realization of comprehensiveness in landscape design on new land be better observed and studied than here in The Netherlands. The Haarlemmermeer Polder (1852) warded off the threat of flood, but it was exploited for the profit of private speculators who subdivided and sold the new land. It represents a typical narrow purpose realization of the "steppe of culture" on new land: a dull system of rectangular parcels and intersecting roads with villages grown up haphazardly around the four corners of the road crossings, and ribbon developments along the main road. It is a landscape of desperate visual monotony; time has not healed the marks of the speculator.

The Wieringermeer Polder represents a transitional stage of planned development. As a government enterprise, it was not undertaken for monetary profit and full consideration was given to the agro-technical aspects of development, including the environmental conditions of farming, such as farmsteads, roads, schools, churches etc. Social planning and landscape design, however, lagged behind and the development was not carried out according to pre-planned stages. The regional scheme created no clear structure and due to the lack of a landscape design the polder has a somewhat nondescript appearance.

Now the North-East Polder is, to the best of my knowledge, the most comprehensively conceived and realized modern enterprise of regional development on new land. Unique success has been achieved in the Northeast Polder by the co-ordination of economic, technological, agricultural, social and physical planning and an exciting new pattern of a regional landscape has been created. By means of an ingenious development scheme in stages, a new, coherent region was created according to plan. These very facts impel one to investigate why this man-made landscape still has a somewhat rigid character. The discussion of some planning problems of the Northeast Polder might bring us nearer to the answer and might further landscape design on both new and old land in many parts of the world. I would like to add that most of the points I am going to mention are in the nature of questions and no more.

The soil survey of the Northeast Polder indicated seven different kinds of soil, but the natural boundaries between soil types were almost completely ignored in the subdivision scheme, and the roads and even the forests are laid out on the gridiron pattern. This is especially obvious in the layout of forest planting which is also done on the gridiron pattern. The general shape of the polder was determined by the intention to include in it the maximum of high quality soils and the minimum of bad soils, and by the need to build the dikes on favourable foundation ground. The shape is not unsatisfactory in general, but it may be that were due consideration taken of the use of landscape as a habitable and hospitable environment both for the local people and for the temporary population of visitors, it is questionable whether this would have been the final shape arrived at. The land drainage system which makes possible the regulation of the ground water level has largely determined the layout of the polder. It is doubtful whether a convincing synthesis has been achieved between the hydrological requirements and the rest of the functional requirements of the regional scheme. Would such a synthesis have been unjustifiable from the economic point of view? I question whether the needs of permanent soil protection and conservation were fully taken into account in the planning of field units, plantations, green strips and so on, and whether such needs have not been overridden by short-range economic and technological considerations. The layout of road profiles and roadside plantings in the polder has created some monotony in the landscape as a whole as well as on the roads themselves. Better ressults could have been achieved by a more variegated planting and less uniform spacing of trees and by the widening of the roads or the roadside plantations at chosen points. Different treatment of the roadsides is still not enough. Are the roads designed, I would like to ask, with the aim of providing the fastest communication channels between settlements, or also with a view to the traveller's interests and delight in observing and experiencing the man-made polder landscape? Rural and urban development in the polder is tailored for an exclusively agricultural region; it is conceivable that a greater diversification of functions – such as more non-agricultural industrial development, or recreational development and consequently an increase in the size of the central town of Emmeloord—should be sought. This would lead with the changed requirements of population, industry, transportation, etc., to a change in the function of the landscape, and this in turn would exert a beneficial influence on the polder landscape pattern as a whole.

The next step in the development of new land is now being planned and realized by the Dutch in the East Flevoland Polder. There is still little information available and the plan I am showing here is not the final scheme but only one of the stages or, rather, a temporary proposal, let us call it. As there is still little information about this project, it is much too early to draw conclusions, but the new trends in its design, as emphasized in the preliminary reports of the planning authorities, are highly significant. I would like to mention particularly the treatment of the roadsides – especially the southeastern road of the East Flevoland Polder, which is a road leading along a dike. This road is not treated as in former projects, for on the water side, to the east, there is an ever-changing strip of landscape of undefended foreland left for recreational purpose, and the great success of this method is visible even today in the first stages of development.

In general this scheme of the East Flevoland Polder reveals a much higher degree of adaptation to the soil map, topography, coastlines and to an overall regional and country-wide plan than any of the previous polder projects. Three types of farms will be established according to soil characteristics, and the polder is being designed with a view to the total needs of the Dutch community and not those of only one sector. In a recent publication of the Dutch Ministry of Public works, the purpose of the planning of this new land is stated as: the creation of a country which is not only suitable for work but which is also a good place of habitation. The distribution of villages and townships in the new polder has been – at least in the plan – adjusted to recent experience with the influence of modern communications and methods of cultivation in village development. For the purpose of providing the rural population with a high standard of services, the number of villages has been reduced and distances and scales increased accordingly in a natural way. The plan also provides for a central town of 30,000 to 60,000 inhabitants, with good connections with the centre of The Netherlands and with an industrial development independent of local agricultural production. The importance of planning for the national recreational needs has this time been recognized as a primary requirement. In conjunction with technical requirements a lake is being created as well as a partially afforested, un-fenced foreland beach. Recreational areas have also been planned near the villages and throughout the polder. It therefore seems that the progress of landscape design on new polderland to be achieved in the East Flevoland Polder will not be less important and exemplary than that of the earlier project.

X

THE INTEGRAL HABITATIONAL UNIT

In planning new residential quarters it should be realized that there is all the difference between building a group of houses which constitutes administratively a "Housing Scheme," and the development of an environment which deserves the name: a town, or a town quarter. Our problem is how to create a valuable and permanent urban environment in a period of urgent demands for popular housing and of mass-production methods. In many countries – and Israel is certainly one of them – the mass construction of houses, which came as a response to waves of population influx into the towns, has led to various disappointments. All too often new residential quarters are not only esthetically embarrassing – being monotonous and characterless – but also socially and economically unmanageable. In some of the new housing schemes a constant flow of inhabitants to other parts of the town is apparent. As a result the less energetic and enterprising elements of the population are left behind and the new quarter gradually deteriorates. In many housing schemes the dwellings, the distances between houses and various services may be more convenient than those in the "old town," but the quarter as a whole is uniform and unsatisfactory. In some cases new housing schemes have "aged" too soon and turned into a sort of old people's quarters, because they were planned for and populated by families of only one type and age. In the popular mind, and unfortunately among many housing specialists, also, "housing" is still considered as the creation of an environment adapted to the requirements of the so-called "average family." This approach cannot yield results which will satisfy communities for a considerabble length of time.

To plan a residential urban quarter one must first understand what a town as a place of living and working truly is. Towns are the centres of regional communities, and are the organs by which these regions unite and come into contact with the larger national community and with humanity. The values of living in towns derive from the varied opportunities for self-expression and intercourse they offer. Within a strictly limited area there agglomerate various forms of economic activity, of social interaction between different communities, professions and classes, of cultural roles and of buildings and squares with different functions. Whereas the village is characterized by the parallelism of the functions of its "cells"—the farming families—the town thrives on diversification, competition and the integration of contrasting elements. In the course of the attempts to balance or relate the diverse elements and factors to each other, life gains a new dimension and becomes *urban life*. The creation of such must be fostered also in the contemporary town.

Urban equilibrium was achieved in the past through a long process of trial and error at a time when the urban population constituted only a fraction of the total regional population. In the present period, the steadily increasing flow into towns in all developing countries, and the rising economic and cultural importance of the town in the life of countries, have made it impossible. The unplanned growth or multiplication of urban areas involves unbearable social costs and economic chaos. At the same time it must be taken into consideration that where a composite urban environment is emerging in a continuous historic process, the chances of inflicting esthetical mischief and unhuman rigidity are much smaller than in places where an environment is created "in a single stroke," in a planned manner. Yet towns or town quarters must be built nowadays within a short span of time. The solution of this problem, and the creation of an habitable urban environment, can be achieved only by the utmost application of our scientific and artistic capabilities.

The planning of the Integral Habitational Unit is meant as an effort of this kind; although it does not solve the problem of the contemporary town as a whole, it may represent progress as regards the residential units of which a town is composed. The idea originated with Le Corbusier, who coined the term *unité d'habitation*, although he applied it only to his multi-storey blocks; the idea was elaborated and developed by the Municipality of Rotterdam and American universities. In Israel, we are planning an experimental unit of this type in Kyriat Gat. It is obvious that the Integral Habitational Unit would be of special interest to a country like Israel, where community integration is a national aim.

The Integral Habitational Unit is an attempt to fulfill the most essential urban requirement: Unity in Variety. It is an attempt to overcome the regimentation of population and the uniformity of mass housing by creating mixed residential units, representative of a wide range of variations in the population. The idea is based on the belief that the emergence of the values of urban life can be assisted by deliberate environmental creation. By shaping a suitable relationship between the various individual dwellings of the different types of the urban population, and by relating them to the community services and centres, we hope to foster urban community development and arrive at new composite urban structures.

In planning and designing the Integral Habitational Unit, it is essential that the planners should always perform a dual function: a study and a taking into account of a maximal variety of individual housing requirements, and the integration of the great variety of different homes into a single residential unit. In performing the first function, we must rid ourselves of the misleading notion of using the average family as a basis for a housing program. In Israeli development towns we have come to distinguish five "dimensions" which serve as measuring rods of the varying housing requirements of the population: the size of the family, the age of the head of the family, the ethnic community, the occupation and economic level and the length of time the family has lived in the country. Such studies must be done separately for each project and place by demographers. On this basis a number of housing types answering the variety of requirements realistically can be prepared: the number of dwellings, according to kind and size, can be roughly estimated. Wherever possible, such programs should be planned for long range requirements. The housings program of the Integral Habitational Unit is therefore planned with reference to a population projection, which, for the sake of reliability, does not exceed ten years. In the experimental unit of Kiryat Gat, numbering 900 dwellings, we have proposed 15 different basic types of dwelling.

Once the types of dwellings required in the Integral Habitational Unit have been determined, the optimal relationships between them should be established with the aid of sociological research. New urban quarters built in Israel are mostly of heterogeneous composition as far as ethnic origins are concerned. Ethnically homogeneous neighbourhoods are not only undesirable from the national point of view, but also have been often socially unsuccessful, as proved by research. But a completely unplanned, haphazard mingling of families has led to unsatisfactory neighbours' relations and did not enhance the feeling of belonging to a place. The optimal composition of a neighbourhood has therefore still to be discovered. Sociological research is slowly making progress towards answers to these questions. We have learned that such research must be stretched over several years in order to yield valuable practical results. Even then its final aim might not be to find the best "prescription" for the social composition of a residential quarter, but to elucidate the factors which exert a favourable or unfavourable influence on the development of relationships within a community and between a community and its environment. In the sociological research of housing problems, it is of the highest importance to arrive at a continuous collaboration between the architect-planner and the scientist. These two professions, unfortunately, still think and work in different terms and with different aims in their minds.

For planning the Integral Habitational Unit, the local economic survey and plan will be important in various ways: it sheds a light on the present and future professional make-up of the population, making it possible to plan an Habitational Unit truly representative of the town's economic structure; it helps to relate the incomes of the various groups of the population to the cost of houses to be rented or purchased; it establishes the kinds and number of shops and other services which a specific group of residents will support; it is of great help in arriving at decisions on the building methods to be employed on the site, depending on the skill and availability of labour, the distances, to centres of industrial production, the local industrial potential, etc. Besides, the economic research can help the planner greatly in a general way by making him familiar with many material problems of the people for whom he is planning.

In order to define the size and structure of an Integral Habitational Unit, we follow several hypothetical principles: the Unit should be large enough to support a small but variegated group of shops and possibly a cultural centre; it should more or less suit the absorptive capacity of one two-stream elementary school (16 classes); the physical extent of the Unit and the location of its centre should render possible easy pedestrian communication within the quarter; it should form for its inhabitants an easily surveyable architectural whole with a clearly defined nucleus - the services' and social centre.

We proposed that the size of the Integral Habitational Unit be related to a two-stream elementary school, which should be supported by about 800 families. Its sub-units or "cells" can also be determined by following the optimal absorptive capacity of secondary services such as one-class or two-class kindergartens, which may serve 250 or 500 families respectively, or a playground for children between 5 - 11 years of age serving 200 families each, or a small and well-defined open space which in combination with the playground would form the primary recreational space for a group of 200-250 families. Theoretically, therefore, an Integral Habitational Unit would contain about 800 families and might be composed of three to four sub-units of 200-250 families each. As these figures should be related to the demographic social, educational, geographical and topographical conditions varying widely in each locality, they can only be taken as approximations.

By determining the dimensions, interior distances, density and location of the community "nucleus" in an Integral Habitational Unit, the privacy of the home is brought into an harmonious relationship with the public nature of the Unit's open spaces, which will be planned for a desirable frequency of personal contacts. The Integral Habitational Unit should constitute a convenient "shell" for its inhabitants and at the same time an instrument of social integration – at least in those countries where housing problems are closely connected with problems of social change.

As designers we consider it essential to create environmental values and to overcome the danger of building nondescript towns and quarters. In this task we are greatly assisted by scientific research, which helps us to clarify the particular problems of a location and a community in statu nascendi. In this manner a diversified and realistic program becomes the basis of integrated planning. Our cooperation with scientists and community-workers is therefore vitally important – but it is not sufficient to guarantee the succes of our work. In the final analysis the creation of a satisfactory environment is a work of art. It requires an intensive process of design, and this means not only the beautification of the technically needed structures and spaces of a town. The transformation of the scientifically and administratively defined necessities into a design of distinct quality and character is an independent process within the web of operations needed to create a new town or town-quarter.

An Integral Habitational Unit is transformed into a unified plastic creation by a three-dimensioned design. It must be integrated with the town of which it is a part and related to its surrounding landscape. The unit must be visualized as an object seen from various viewpoints in the regional landscape, or through the eyes of the approaching traveller. Its skyline must be related to the background's skyline, whether mountainous or otherwise. But the landscape, as observed from within the unit, must also become part of the inner-urban environment, and should be studied as an horizon of varying depths and local points. From the Unit vistas may be opened on interesting landscape features, while being closed in other directions. For such a design the survey of landscape horizons as they evolve around a town or a town-quarter might be of great value.

The sense of the Integral Habitational Unit as a unified whole will be brought to its system of interior spaces – the transition from the enclosed spaces between buildings to wider spaces, then to open spaces, which is experienced when walking in the Unit or travelling through it. A whole scale of human needs, from the need for individual seclusion to the need for gregariousness, can be satisfied by the design of a system of interlocking space. Thought about interior urban space become most important in our time because of the tremendous change effected in the reality as well as in our conception of the town by motorized transportation. Instead of one system of communications in towns, as in former times, we now need two in order to escape chaos and danger and to offer the citizen proper space for movement and rest. Driven away from the street by the motorcar, the citizen is regaining his right to free unhampered movement and social contact in the open within the interior urban spaces provided by modern planning.

This "retreat" from the street, along with the new discovery of interior urban space, the recognition of social requirements and modern construction methods, open new and interesting possibilities to urban design. Yet we are still far from a new and comprehensive conception of the new town. In the traditional towns which we rightly admire, the balanced and unified urban form grew from a central, social, political or religious idea. Our own dual, scientific-artistic, approach to the problems of town development is in fact only a compensation for the lack of such central themes. Honest definitions of the purposes of contemporary urban design, therefore, are characterized by a measure of modesty. Probably this approach cannot lead to masterpieces of urban design. Nevertheless there is scope for the creation of the basic elements of which the contemporary and the future town must be constituted. Herein lies also the importance of the comprehensive detailed planning of the Integral Habitational Unit. We shall certainly have to revise in the future many ideas on which the Integral Habitational Units are based. But no effort should be spared to introduce into the plans of new residential quarters the true measures of their social structure, and to establish relationships in the lay-out which foster the integration of emerging communities.